BRYAN MULLEN,

42, ERNEST STREET,

MERTHYR TYDFIL.

Applied Heat

APPLIED HEAT

T. H. Thomas, M.Sc.(Eng), C.Eng,
M.I.Mech.E., M.I.Prod.E., A.M.I.Struct.E.
Formerly Vice-principal and Head of Mechanical and Production
Engineering Department, Chesterfield College of Technology

and

R. Hunt, B.Eng, C.Eng, A.M.I.Mech.E.,
A.M.I.Mar.E.
Head of Department of Mechanical Engineering, Poplar Technical
College

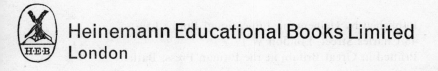 **Heinemann Educational Books Limited**
London

Heinemann Educational Books Ltd
LONDON MELBOURNE TORONTO SINGAPORE
CAPE TOWN AUCKLAND IBADAN HONG KONG

Published by Heinemann Educational Books Ltd
48 Charles Street, London W.1
Printed in Great Britain at the Pitman Press, Bath

Contents

Contents

Preface

This text is written with the aim of presenting the elements of applied thermodynamics in as fundamental a manner as possible. The approach is a simple one, bearing in mind that the book is intended primarily as a first text for the second year of the National Certificate in Engineering and for the first year of Diploma courses.

It is modelled on the energy concept pioneered so successfully by Keenan in America. The idea of systems enclosed by boundaries has been emphasised throughout, and experience has shown that students taught on these lines have gained a considerable confidence. Some pains have been taken to eliminate the false concept that heat can be stored. Heat is presented as a transient form of energy present when a temperature difference exists. When the implications of this method are realised students have no difficulty with internal energy and the more difficult concepts.

Syllabuses vary from college to college and from one examining body to another, and the authors were tempted to add further chapters but resisted the temptation in the interests of compactness.

It will be noted that descriptions of the 'hardware' of heat engines have been excluded. It was felt that most students have access to college engineering laboratories where they can see such 'hardware' at first hand; certainly as much as would be applicable to a book of this kind.

The symbols used in the text conform as far as possible to those recommended in B.S. 1991. Relevant exercises for the student will be found at the end of chapters.

List of Symbols

Symbol	Quantity	Units
A	Surface area	ft^2 or in^2
c	Velocity	ft/sec
c_p	Specific heat at constant pressure	Btu/lb degR or Btu/lb degK
c_v	Specific heat at constant volume	Btu/lb degR or Btu/lb degK
g	Gravitational acceleration	ft/sec^2
g_c	Gravitational constant	$lb\ ft/lbf\ sec^2$
H	Enthalpy	Btu
h	Specific enthalpy	Btu/lb
k_d	Diagram factor	
L	Latent heat	Btu/lb
l	Length	ft or in
m	Mass	lb
n	Polytropic index	
P or p	Pressure	lbf/in^2
Q	Heat transfer	Btu
R	Characteristic gas constant	ft lbf/lb degR or ft lbf/lb degK
R_0	Universal gas constant	ft lbf/lb Mol degR or ft lbf/lb Mol degK
r	Expansion ratio	
r_v	Volume compression ratio	
r_p	Pressure ratio	
S	Entropy	
s	Specific entropy	
T	Absolute temperature	°R or °K
t	Temperature	°F or °C
U	Internal energy	Btu
u	Specific internal energy	Btu/lb
V	Volume	ft^3
v	Specific volume	ft^3/lb
W	Work or	ft lbf
	weight	lbf
Z	Height above a datum	ft
η	Efficiency	
γ	Adiabatic index	

1: Introduction

1.1 Introduction

Applied heat is a study of the relationships between heat energy, work, and the properties of matter, and also of the conversions of heat energy and work by means of heat engines. These heat engines employ an elastic fluid as their working medium and it is this fluid which makes possible the energy conversions. It is important to define certain concepts precisely if a thorough understanding is to be obtained. These are discussed below.

1.2 System

A system may be defined as a certain collection of matter within an arbitrary prescribed region, e.g. in Fig. 1.1, the fluid contained by the

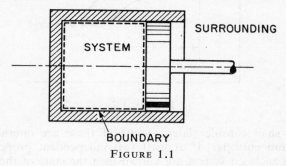

FIGURE 1.1

cylinder head, cylinder walls and the piston may be said to be the system.

1.3 Surroundings

The region enclosing the system is designated the surroundings. The surroundings may be affected by changes within the system.

1.4 Boundary

This is the surface of separation between the system and its surroundings. It may be the cylinder and the piston as in Fig. 1.1 or an

1

imaginary surface drawn so as to enable an analysis of the problem under consideration to be made. Systems are further sub-divided into closed systems and open systems. In a closed system the fluid remains within the boundary. In an open system, in addition to heat energy and work energy crossing the boundary, the fluid also is allowed to cross the boundary.

1.5 Properties

The properties of a system are those characteristics which define its condition. The most common of these in thermodynamics are pressure, volume and temperature, although there are many other useful ones

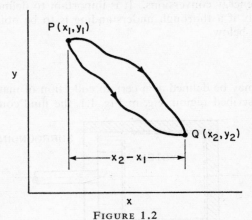

FIGURE 1.2

which we shall consider later. Some of these are internal energy, enthalpy and entropy. If at least two independent properties of a fluid in an enclosed system are known then the state of the system is known, and the condition of the system is completely described. This is similar to plotting a point P on a graph having two coordinate axes x and y. If the coordinates of P are known, say x_1 and y_1, then the position of P on the graph is completely described. Thus, if x_1 and y_1 (i.e. point P) define the state of a system at one instant whilst x_2 and y_2 (i.e. point Q) define its condition at a later instant of time, then these points can be plotted on a graph as shown in Fig. 1.2. Regardless of the path through which the system passes in going from P to Q it can be seen that $x_2 - x_1$ is a fixed quantity which is entirely independent of the path. It is important to remember that a property has this fundamental characteristic, i.e. that the change in its value is independent of the path traversed in going from the initial to the final state. If a description of a quantity which does not meet this requirement then the quantity is not a property. It will be seen later that work transfer and heat transfer are not properties; because of this, work transfer is not written as $W_2 - W_1$ or heat transfer as $Q_2 - Q_1$ in a process. Instead, they must be written W_{12} (or simply W) and Q_{12} (or Q). The two very important properties of pressure and temperature, are discussed in Appendix 1.

1.6 Thermal Equilibrium

If a system is completely stable then one particular value of any property will be the same at all points throughout the system at that particular instant. Thus, for a system to be stable no energy transformations (e.g. by heat transfer, work transfer, eddying, etc.) must

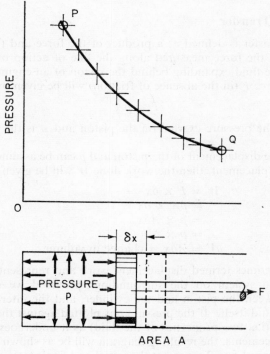

FIGURE 1.3

be taking place. Under these conditions a system is said to be in thermal equilibrium and one particular value of a property may then be ascribed to the system as a whole. If the values of two independent properties are known then the state of the system can be represented as a point on a graph of its properties.

When the properties of a system change it is usual to assume that the initial and final conditions are states of thermal equilibrium enabling them to be plotted on a graph of the properties. Between the initial and final conditions, any other equilibrium states may also be plotted as definite points on the graph as shown in Fig. 1.3. If the process takes place in a series of infinitely small steps, and after each step the system is allowed to settle in thermal equilibrium, then the complete process will appear on the graph as a continuous line. At all points during the process the state of the system would be completely known. Such a process, called a reversible process, would require an infinitely long time to complete. In actual practice, it is impossible to realise a reversible process as all real processes must take place in a definite time. These real processes are known as irreversible processes.

1.7 Work Transfer

Work transfer is defined as a product of the force and the distance travelled by the force measured along the line of action of the force. Consider the fluid expanding behind the piston of an engine as in Fig. 1.3. The force F (in the absence of friction) will be given by

$$F = pA$$

where p is the pressure exerted on the piston and A is the area of the piston.

If δx is the displacement of the piston and p can be assumed constant over this displacement, then the work done W will be given by

$$\begin{aligned} W &= F \times \delta x \\ &= pA \times \delta x \\ &= p \times A\delta x \\ &= p \times \delta V \end{aligned}$$

where $\qquad \delta V = A\delta x =$ change in volume

This is sometimes termed displacement work and represents the work done by the system on the surroundings provided we neglect the friction between the piston and the cylinder, and the internal friction within the fluid itself. If the pressure p is plotted against the volume V of the fluid (i.e. two properties of the fluid) as it undergoes a series of small displacements, the resulting diagram will be as shown in Fig. 1.4. In this diagram 1 and 2 represent the initial and final states and we may plot a continuous line from 1 to 2 through B representing the succession of states. The work done under the above assumption of absence of friction effects will be equal to the sum of $p\delta V$ for each small step between 1 and 2. This is usually written in the form $\Sigma_1^2 p\delta V$. It can be seen from Fig. 1.4, that this will be represented by the shaded area. If the line joining the series of successive states is along a different path (say through C) then it is obvious that the area (and therefore the work done) will have a different value. This shows that the work done (or work transfer) is not a property of the fluid since it is not independent of the

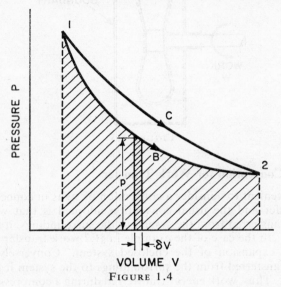

FIGURE 1.4

path along which the fluid passes to its final state. Again, it must be noted that the fluid does not possess any work but that the work energy transfer may only take place when there is a change of volume from 1 to 2.

1.8 Paddle or Stirring Work Transfer

It was seen in the last section that work was capable of being transferred from the system to the surroundings by a volume change, and that $\Sigma_1^2 p \delta V$ gave a measure of the work provided that friction forces were zero and that pressure equilibrium prevailed in the system.

Consider a paddle being rotated in a fluid by the action of an external torque applied to the paddle shaft. It is clearly seen that there will be work transfer although there will be no volume change (see Fig. 1.5). Here, therefore, the value of $\sum p \delta v$ will not be a measure of the work transfer since $\delta V = 0$. The work transfer will take place by means of the friction forces acting at the surface of the paddle, and must be measured by a knowledge of other properties of the system. If the frictional forces in the system are reduced, the external torque and, consequently, the work transfer will also be reduced. With a frictionless fluid the work transferred by means of paddle action is zero.

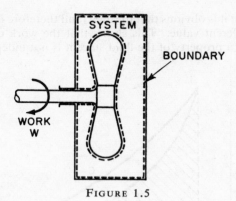

FIGURE 1.5

1.9 Sign Convention for Work Transfer

It is convenient to consider a convention of signs in connection with work transfer and the usual convention adopted is that when work energy is transferred from the system to the surroundings, it is denoted as positive. In the case of the example in §1.7 work transfer is positive during the expansion of the enclosed system. Conversely, if work energy is transferred from the surroundings to the system it is denoted as negative. Thus, work energy transferred during a compression would be negative.

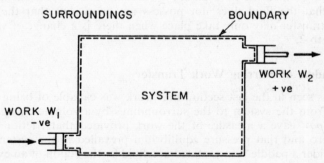

FIGURE 1.6

1.10 Heat Transfer

Heat is a form of energy which crosses the boundary of a system during a change of state produced by a difference of temperature between the system and its surroundings. Referring to Fig. 1.4 where a system changed its state along different paths, it was seen that the amount of work energy transferred depended upon the path taken.

Similarly it can be shown—but not so easily as in the case for work transfer—that the heat transfer Q also depends upon the path taken. Q therefore is not a property of the system. It may be defined as energy in transference due to a temperature difference and it is not a quantity which can be stored as was conceived by early investigators 2–3 centuries ago.

1.11 Sign Convention for Heat Transfer

The sign convention usually adopted for heat energy transfer is such that if heat energy flows into the system from the surroundings it is said to be positive. Conversely, if heat energy flows from the system

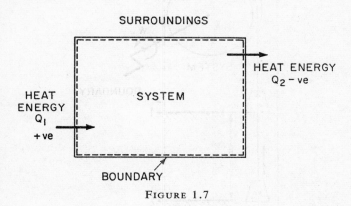

FIGURE 1.7

to the surroundings it is said to be negative. This convention is illustrated in Fig. 1.7. It is incorrect to speak of heat in a system since heat energy exists only when it flows across the boundary. Once in the system, it is converted to other types of energy.

1.12 Unit of Heat

The unit of heat is taken as the amount of heat energy required to raise the temperature of unit mass of water through 1 degree whilst under standard atmospheric pressure. If the unit of mass is taken as 1 pound and the temperature change 1 degree Fahrenheit, the unit of heat is one British Thermal Unit (1 Btu). For the same mass of 1 pound and a temperature change of 1 degree Centigrade (or Celsius), the unit of heat is the Centigrade Heat Unit (Chu). Much more exact definitions have been made but the above are sufficiently accurate for engineering purposes.

1.13 The First Law of Thermodynamics

Consider an ideal frictionless system Fig. 1.8 consisting of a weight w connected to a drum by means of a rope. When the system changes from state 1 to state 2 (i.e. the weight lifted from plane 1 to plane 2), a certain amount of work $W = w \times h$ must cross the boundary into the system, where it is converted into potential energy. This quantity of work energy W will have a negative sign since the work transfer takes place from the surroundings into the system. If the system is now

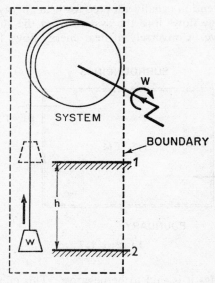

FIGURE 1.8

allowed to return to its original state (that is, if the weight w is allowed to return to its original plane 1 slowly by applying a restraining force to the crank) the change in potential energy wh will be converted into work energy W which will return from the system to the surroundings across the boundary. This work will be positive in sign and it follows that the nett work transfer will be zero. This may be written $\sum \delta W = 0$, where $\sum$ means the sum of all the little bits of work δw in going from 1 to 2 and back to 1. It is interesting to note that we do not say that work is stored at 2 but that energy is stored at 2 which is termed potential energy. Since the system has returned to its initial state after undergoing a series of processes, it is said to have undergone a closed cycle.

Suppose the process of lifting the weight from 1 to 2 is repeated and as the weight returns from 2 back to 1 let the motion be controlled by means of a brakeband as in Fig. 1.9. The work transfer during the descent of the weight will now be converted into heat energy Q by the brakeband. If the heat energy Q is accurately measured for weights moving through various heights, we find that the values of work transferred are directly proportional to the heat energy measured, i.e. $W \propto Q$ or $\sum \delta W \propto \sum \delta Q$. If the sign of proportionality is replaced by an equality sign then $\sum \delta W = \text{constant} \times \sum \delta Q$. This constant is

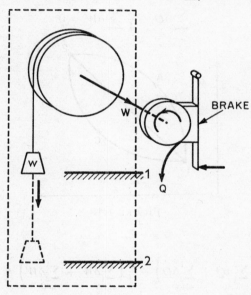

FIGURE 1.9

given the symbol J after Joule, who performed some classical early experiments in this field. J is called the mechanical equivalent of heat. If Q is 1 Btu and W is the work done in ft lbf, then W is equal to J ft lbf. The value of J in this case is accepted as 778 ft lbf. In centigrade units, the corresponding value is J = 1,400 ft lbf.

The equation

$$\sum \delta W = J \sum \delta Q$$

or

$$\frac{1}{J} \sum W = \sum Q$$

represents one form of the First Law of Thermodynamics. It may be stated in words as follows: 'If a system is taken through a *closed* cycle, then the nett work energy transferred is directly proportional to the nett heat energy transferred.' There is no mathematical proof of this law but all experimental work supports it.

1.14 Internal Energy

Consider a change in the system from a state 1 to a state 2 by path A and return to the initial state 1 by path C, as illustrated in Fig. 1.10. As a consequence of the First Law,

$$\sum \delta Q - \frac{1}{J} \sum \delta W = 0 \tag{1.1}$$

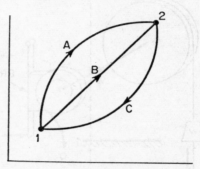

FIGURE 1.10

or

$$\left(\sum_1^2 \delta Q + \sum_2^1 \delta Q \right) - \frac{1}{J} \left(\sum_1^2 \delta W + \sum_2^1 \delta W \right) = 0 \tag{1.2}$$
via A   via C   via A   via C

or

$$\left(\sum_1^2 \delta Q - \frac{1}{J} \sum_1^2 \delta W \right) + \left(\sum_2^1 \delta Q - \frac{1}{J} \sum_2^1 \delta W \right) = 0 \tag{1.3}$$
via A   via A   via C   via C

Now suppose that the system changes from state 1 to state 2 via the path B and returns to state 1 by the path C. By a similar reasoning

$$\left(\sum_1^2 \delta Q - \frac{1}{J} \sum_1^2 \delta W \right) + \left(\sum_2^1 \delta Q - \frac{1}{J} \sum_2^1 \delta W \right) = 0 \tag{1.4}$$
via B   via B   via C   via C

Comparing equations (1.3) and (1.4), it is seen that

$$\left(\sum_1^2 \delta Q - \frac{1}{J} \sum_1^2 \delta W \right) = \left(\sum_1^2 \delta Q - \frac{1}{J} \sum_1^2 \delta W \right) \tag{1.5}$$
via A   via A   via B   via B

Therefore,

$$\sum_1^2 \delta Q - \frac{1}{J}\sum_1^2 \delta W$$

has the same value irrespective of the path taken from state 1 to state 2, and must, therefore, represent a change in the property of a system. This property is termed Internal Energy, denoted by U, and hence the change in internal energy between states 1 and 2 is given by

$$U_2 - U_1 = \sum_1^2 \delta Q - \frac{1}{J}\sum_1^2 \delta W$$

$$= Q_{12} - \frac{W_{12}}{J} \tag{1.6}$$

Internal energy is the sum of all the energies the fluid possesses and stores within itself. The molecules of a fluid may be imagined to be in motion thereby possessing kinetic energy of translation and rotation as well as the energy of vibration of the atoms within the molecule. In addition the fluid also possesses internal potential energy due to inter-molecular forces. The total of these energies contribute largely to the internal energy. When a gas expands without heat supply from a high to low pressure behind a piston, it is the internal energy which enables the gas to do work. In fact in this case the change of internal energy $U_2 - U_1$ equals the work done (W/J). It is impossible to determine the absolute value of the internal energy at any one state, but this is no disadvantage since we invariably require to know only the change of internal energy. The internal energy of a unit mass is termed the specific internal energy u and the total internal energy $U = mu$, where $m =$ mass of fluid.

Ex. 1.1. Figure E.1.1 shows a certain process which undergoes a complete cycle of operations.

Determine the value of the output work W_0.

For a complete cycle

$$\sum Q - \sum W = 0$$

Here Q_i is +ve, W_0 is +ve, Q_0 is −ve, W_i is −ve

$$\therefore \qquad \sum Q = 10 - 3, \quad \sum W = W_0 - 2$$

Hence
$$(10 - 3) - (W_0 - 2) = 0$$
$$7 - W_0 + 2 = 0$$
or
$$\underline{W_0 = 9 \text{ Btu}}$$

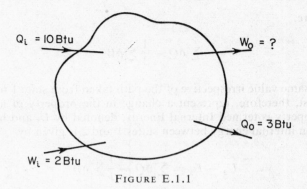

FIGURE E.1.1

Ex. 1.2. A system is allowed to do work amounting to 389,000 ft lbf whilst heat energy amounting to 800 Btu is transferred into it. Find the change of internal energy and state whether it is an increase or decrease.

$$U_2 - U_1 = Q_{12} - \frac{W_{12}}{J}$$

now

$$\frac{W_{12}}{J} = + \frac{389,000}{778} = + 500 \text{ Btu}$$

Hence $U_2 - U_1 = 800 - 500 = 300 \text{ Btu}$

Since $U_2 > U_1$, the internal energy has increased.

EXERCISES ON CHAPTER 1

1. During a complete cycle of operations a system is subjected to the following heat transfers; 800 Btu supplied and 550 Btu rejected. At two points work is done by the system to the extent of 70,000 ft lbf and 15,000 ft lbf. At a third point there is a further work transfer. Determine its amount and state whether it is work done by or on the system.

       (Work = 109,000 ft lbf and is work done by the system)

2. An engine dynamometer absorbs 120 hp during a trial on an engine. The heat generated in the dynamometer is carried away by cooling water. After steady conditions have been obtained determine the quantity of cooling water flowing for a temperature rise of 30 degF.

       (169 lb/min)

3. A fan runs in a closed chamber. At a point when its kinetic energy is 500,000 ft lbf, the power to the fan is cut off and the fan gradually runs down to rest. It is estimated that the heat loss from the chamber during the time the fan runs down is 50 Btu. Determine the change of internal energy of the contents of the chamber.

       (593 Btu)

4. In an oil engine trial it is found that the energy rejected to the cooling water, to the exhaust gases, and to the surroundings amount to 30, 40 and 8% respectively. The useful work developed by the engine is 10 hp. Determine the necessary heat input in Btu/min and the amount of fuel oil required per hour if the heat evolved per lb of oil is 18,000 Btu. Draw a line diagram of the engine showing the energy flow.

(1,930 Btu/min, 6·43 lb/h)

5. In an experiment to determine the mechanical equivalent of heat a drum having a total water equivalent of 0·09 lb is driven against a brake which applies a braking torque of 2·71 lbf in. After the brake drum has turned through 110 revolutions the corrected rise in temperature of the water and drum is found to be 2·231 degF. Determine the value of the mechanical equivalent of heat.

(781 ft lbf/Btu)

6. If the pressure behind a piston remains constant at 100 lbf/in² while the volume increases from 0·1 ft³ to 0·8 ft³ and heat amounting to 6 Btu is radiated from the cylinder walls, determine the change of internal energy of the contents.

(−18·9 Btu)

7. State which of the following are properties of a fluid: pressure, temperature, volume, heat transfer, work. Give the reasoning in support of your answer.

2: Steady Flow Processes

The means by which changes in the state of a thermodynamic system are effected are called processes. These processes may be broadly divided into two types, depending on whether the system is an open or a closed system.

2.1 Non-flow Processes

In a closed system, although energy may be transferred across the boundary in the form of work energy and heat energy, the working fluid itself never crosses the boundary. Any processes undergone by a closed system are referred to as non-flow processes.

2.2 Flow Processes

In an open system, in addition to energy transfers taking place across the boundary, the fluid may also cross the boundary. Any process undergone by an open system is called a flow process. These processes may be sub-divided into unsteady flow processes and steady flow processes.

2.3 Steady Flow Processes

In heat engines it is the steady flow processes which are generally of most interest. The conditions which must be satisfied by all of these processes are:

  (i) The mass of fluid flowing past any section in the system must be constant with respect to time
 (ii) The properties of the fluid at any particular section in the system must be constant with respect to time
(iii) All transfers of work energy and heat energy which take place must do so at a uniform rate

A typical example of a steady flow process is a steam boiler, shown diagrammatically in Fig. 2.1, operating under a constant load. In order to maintain the water level in the boiler, the feed pump supplies water at exactly the same rate as that at which steam is drawn off from the boiler. To maintain the production of steam at this rate at a steady pressure, the furnace will need to supply heat energy at a steady rate.

14

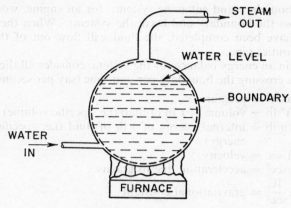

FIGURE 2.1

Under these conditions, the properties of the working fluid at any section within the system must be constant with respect to time.

2.4 Steady Flow Energy Equation

To analyse a steady flow process the fact is used that energy cannot be created or destroyed but only converted from one form to another. The resulting energy balance for a steady flow process is called the

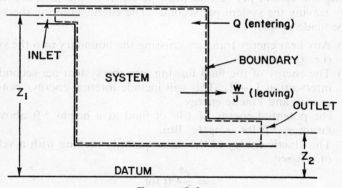

FIGURE 2.2

Steady Flow Energy Equation. Consider a system undergoing a steady flow process (see Fig. 2.2).

The working fluid will flow along the inlet pipe at a constant rate and enter the system, where various energy transfers take place, depending on the function of the system (e.g. for a boiler, heat energy would cross the boundary and enter the system; for an engine, work energy would cross the boundary and leave the system). When these energy transfers have been completed, the fluid will flow out of the system along the outlet pipe.

To obtain an energy balance for the system, consider all the amounts of energies crossing the boundary per unit time (say per second).

Let p lbf/ft^2 = pressure
  v ft^3/lb = volume of 1 lb of fluid (i.e. specific volume)
  u Btu/lb = internal energy of 1 lb of fluid (i.e. specific internal energy)
  c ft/sec = velocity
  g ft/sec^2 = acceleration due to gravity
  $g_c \dfrac{\text{lb}}{\text{lbf}}\dfrac{\text{ft}}{\text{sec}^2}$ = gravitational constant
  z ft = height
  Q Btu/sec = nett energy transfer taking place in the form of heat energy crossing the boundary into the system
  W/J Btu/sec = nett energy transfer taking place in the form of work energy crossing the boundary out of the system
  m lb/sec = rate at which the fluid flows through the system

Let suffix (1) denote conditions at inlet to the system
Let suffix (2) denote conditions at outlet to the system

Applying an energy balance to the system, the total amount of energy entering the system per second must be equal to the total amount of energy leaving the system per second. The energy entering the system will be made up of:

(i) Any heat energy transfers crossing the boundary into the system (i.e. Q Btu/sec)

(ii) The energy of the fluid flowing into the system per second past intersection $1 - 1$. This will include internal energy, potential energy and kinetic energy

The potential energy of 1 lb of fluid at a height z ft above the datum = z ft lbf = $z g/\text{J}g_c$ Btu.

The kinetic energy of the same quantity moving with a velocity of c ft/sec

$$= \frac{c^2}{2g_c} \text{ ft lbf}$$

$$= \frac{c^2}{2g_c\text{J}} \text{ Btu}$$

Therefore the energy of the fluid per lb

$$= u + \frac{zg}{\text{J}g_c} + \frac{c^2}{2g_c\text{J}} \text{ Btu}$$

If the mass flow rate is m lb/sec, this becomes

$$= m\left(u + \frac{zg}{Jg_c} + \frac{c^2}{2g_cJ}\right) \text{ Btu/sec}$$

Therefore at the inlet to the system this energy

$$= m\left(u_1 + \frac{z_1}{J} + \frac{c_1{}^2}{2gJ}\right) \text{ Btu/sec}$$

(iii) In order for the fluid to enter the system at 1–1, it must displace the fluid immediately preceding it which is just inside the system, i.e. the fluid entering the system must transfer work energy into the system. Since the pressure at 1–1 remains constant at p_1 lbf/ft², the amount of work energy transferred to the system per second will be equal to

$$\frac{p_1 \text{ lbf/ft}^2 \times \text{ volume of fluid per second}}{J \text{ ft lbf/Btu}}$$

$$= \left(p_1 \text{ lbf/ft}^2 \times v_1 \frac{\text{ft}^3}{\text{lb}} \times m \frac{\text{lb}}{\text{sec}}\right) \times \frac{1}{J} \frac{\text{Btu}}{\text{ft lbf}} = \frac{mp_1v_1}{J} \text{ Btu/sec}$$

Therefore the total energy entering the system

$$= Q + m\left[u_1 + \frac{z_1}{J} + \frac{c_1}{2gJ}\right] + \frac{mp_1v_1}{J} \text{ Btu/sec}$$

The energy leaving the system will be made up of:

(i) Any work energy transfers crossing the boundary out of the system $\left(\text{i.e. } \dfrac{W}{J} \text{ Btu/sec}\right)$

(ii) The energy of the fluid flowing out of the system per second past 2–2. This will include internal energy, potential energy and kinetic energy, i.e.

$$\left[u_2 + \frac{z_2g}{Jg_c} + \frac{c_2{}^2}{2g_cJ}\right] \frac{\text{Btu}}{\text{lb}} \times m \frac{\text{lb}}{\text{sec}} = m\left[u_2 + \frac{z_2g}{Jg_c} + \frac{c_2{}^2}{2g_cJ}\right] \frac{\text{Btu}}{\text{sec}}$$

(iii) In order for the fluid to leave the system at 2–2 it must displace the fluid immediately preceding it which is just outside the system, i.e. the fluid leaving the system must transfer work energy to the surroundings. Since the pressure at 2–2 remains constant at p_2 lbf/ft², the amount of work energy transferred to the surroundings per second by the fluid leaving the system will be equal to

$$\frac{p_2 \text{ lbf/ft}^2 \times \text{volume of fluid per second}}{J \text{ ft lbf/Btu}}$$

$$= (p_2 \text{ lbf/ft}^2 \times v_2 \text{ ft}^3/\text{lb} \times m \text{ lb/sec}) \times \frac{1}{J \text{ ft lbf/Btu}}$$

$$= \frac{m p_2 v_2}{J} \text{ Btu/sec}$$

The total energy leaving the system per second

$$= \frac{W}{J} + m\left[u_2 + \frac{z_2 g}{J g_c} + \frac{c_2{}^2}{2 g_c J}\right] + \frac{m p_2 v_2}{J} \text{ Btu/sec}$$

Since energy entering system per second = energy leaving system per second

$$Q + m\left[u_1 + \frac{z_1 g}{J g_c} + \frac{c_1{}^2}{2 g_c J}\right] + \frac{m p_1 v_1}{J} = \frac{W}{J} + m\left[u_2 + \frac{z_2 g}{J g_c} + \frac{c_2{}^2}{2 g_c J}\right]$$
$$+ \frac{m p_2 v_2}{J}$$

$$\therefore \; Q - \frac{W}{J} = m\left[\left(u_2 + \frac{p_2 v_2}{J}\right) - \left(u_1 + \frac{p_1 v_1}{J}\right) + \frac{c_2{}^2 - c_1{}^2}{2 g_c J} + \frac{z_2 g - z_1 g}{J g_c}\right]$$

$$(2.1)$$

The combination of $(u + pv/J)$ occurs frequently in thermodynamics. It is given the name of *specific enthalpy* and is usually denoted by h.
  Thus

$$h_1 = u_1 + \frac{p_1 v_1}{J}$$

and

$$h_2 = u_2 + \frac{p_2 v_2}{J}$$

and the equation may be written

$$Q - \frac{W}{J} = m\left[(h_2 - h_1) + \frac{c_2{}^2 - c_1{}^2}{2 g_c J} + \frac{z_2 g - z_1 g}{J g_c}\right] \qquad (2.2)$$

Compared with the other terms, the potential energy term $(z_2 - z_1)/J$ is generally small enough to be neglected, and the equation is normally written

$$Q - \frac{W}{J} = m\left[(h_2 - h_1) + \frac{c_2{}^2 - c_1{}^2}{2 g_c J}\right] \qquad (2.3)$$

This is known as the steady flow energy equation.

2.5 Applications of Steady Flow Energy Equation

The steady flow energy equation may be applied to any apparatus through which a fluid is flowing, provided the conditions stated previously are applicable. Some of the most common cases found in engineering practice are dealt with in detail below.

2.6 Boilers

In a boiler operating under steady conditions, water will be pumped into the boiler along the feed line at the same rate as which steam leaves

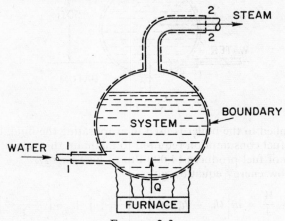

FIGURE 2.3

the boiler along the steam main, and heat energy will be supplied from the furnace at a steady rate. The steady flow energy equation gives

$$Q - \frac{W}{J} = m\left[(h_2 - h_1) + \frac{c_2^2 - c_1^2}{2g_cJ}\right]$$

In applying this equation to the boiler, the following points should be noted:

  (i) Q will be the amount of heat energy passing into the fluid per second

  (ii) W/J will be zero, since a boiler has no moving parts capable of effecting a work transfer

  (iii) $(c_2^2 - c_1^2)/2g_cJ$ will be small compared with the other terms and may usually be neglected

  (iv) m will be the rate of flow of the fluid.

     Hence the equation reduces to

$$Q = m(h_2 - h_1) \tag{2.4}$$

Ex. 2.1. A boiler operates at a constant pressure of 200 lbf/in², and evaporates fluid at the rate of 2,000 lb/h. At entry to the boiler, the fluid has an enthalpy of 80 Btu/lb, and on leaving the boiler the enthalpy of the fluid is 1,185 Btu/lb. The outlet pipe from the boiler is at a height of 50 feet above the inlet pipe, and the inlet and outlet velocities of the fluid are 40 ft/sec and 100 ft/sec respectively. If 65% of the heat

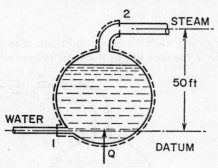

FIGURE E.2.1

energy supplied to the boiler is used in evaporating the fluid, determine the rate of fuel consumption required to maintain this rate of evaporation, if 1 lb of fuel produces 14,000 Btu of heat energy.

Steady flow energy equation gives

$$Q - \frac{W}{J} = m\left[(h_2 - h_1) + \left(\frac{c_2{}^2 - c_1{}^2}{2g_cJ}\right) + \frac{g}{g_c}\left(\frac{z_2 - z_1}{J}\right)\right]$$

Q = heat energy per hour entering system
W/J = work energy per hour leaving system = 0
m = fluid flow rate = 2,000 lb/h
h_2 = 1,185 Btu/lb
h_1 = 80 Btu/lb
c_2 = 100 ft/sec
c_1 = 40 ft/sec
$(z_2 - z_1)$ = 50 ft
g = g_c (numerically)

∴ Steady flow energy equation becomes

$$Q\,\frac{\text{Btu}}{h} - 0 = 2,000\,\frac{\text{lb}}{h}\left[(1,185 - 80)\,\frac{\text{Btu}}{\text{lb}}\right.$$

$$+ \left.\left(\frac{100^2 - 40^2}{2gJ}\right)\frac{\text{Btu}}{\text{lb}} + \frac{50}{J}\frac{\text{Btu}}{\text{lb}}\right]$$

$$= 2,000\,\frac{\text{lb}}{h}\,[1,105 + 0\cdot1678 + 0\cdot06425]\,\frac{\text{Btu}}{\text{lb}}$$

The terms representing the changes in kinetic and potential energy are small compared with the change in enthalpy and may be neglected

$$\therefore \qquad Q = 2,000 \frac{lb}{h} \times 1,105 \frac{Btu}{lb} = 221 \times 10^4 \; Btu/h$$

$$\therefore \text{ heat energy required per hour from fuel} = \frac{221 \times 10^4}{0.65}$$

$$= 340 \times 10^4 \; Btu/h$$

$$\text{heat energy obtained from 1 lb of fuel} = 14,000 \; Btu/lb$$

$$\therefore \text{ fuel required} = \frac{340 \times 10^4}{14,000} \frac{Btu}{h} \frac{lb}{Btu}$$

$$= 243 \; lb/h$$

2.7 Condensers

In principle, a condenser is a boiler in reverse. Whereas in a boiler, heat energy is supplied to convert the liquid into vapour, in a condenser

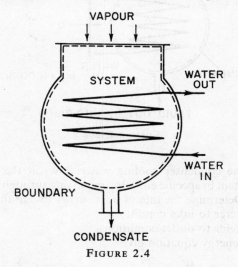

FIGURE 2.4

heat energy is removed in order to condense the vapour into a liquid. If the condenser is in a steady state, then the amount of liquid, usually called condensate, leaving the condenser must be equal to the amount of vapour entering the condenser. The steady flow energy equation gives

$$Q - \frac{W}{J} = m \left[(h_2 - h_1) + \frac{c_2^2 - c_1^2}{2g_cJ} \right]$$

Points to note
    (i) Q will be the amount of heat energy per second transferred from the system
   (ii) W/J will be zero as in the boiler
  (iii) The kinetic energy term may be neglected as in a boiler
  (iv) m will be the rate of flow of the fluid

Thus the equation reduces to

$$Q = m(h_2 - h_1) \qquad (2.5)$$

Ex. 2.2. Fluid enters a condenser at the rate of 80 lb/min with a specific enthalpy of 960 Btu/lb, and leaves with a specific enthalpy of

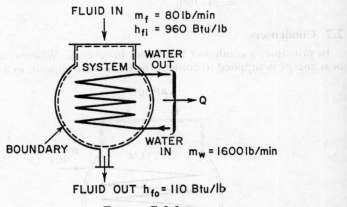

FLUID IN m_f = 80 lb/min
           h_{fi} = 960 Btu/lb

SYSTEM   WATER OUT

         Q

BOUNDARY

WATER IN m_w = 1600 lb/min

FLUID OUT h_{fo} = 110 Btu/lb

FIGURE E.2.2

110 Btu/lb. The condenser cooling water flows at the rate of 1,600 lb/min and its gain in specific enthalpy in passing through the condenser is 40 Btu/lb. Determine the rate of heat energy loss to the atmosphere.
    Let suffix 1 refer to inlet conditions.
    Let suffix 2 refer to outlet conditions.
    Steady flow energy equation gives

$$Q - \frac{W}{J} = m\left[(h_2 - h_1) + \frac{c_2{}^2 - c_1{}^2}{2g_cJ}\right]$$

For a condenser, $W/J = 0$, and the term representing the change in kinetic energy may be neglected. Therefore the equation reduces to $Q = m(h_2 - h_1)$.
    Using this equation, the problem may now be solved by either of the following methods.
    (*a*) Consider the complete system as shown in the diagram. Q

represents the heat energy transfer with the atmosphere. From the above equation

$$Q = m(h_2 - h_1)$$
$$= mh_2 - mh_1$$
$$= \text{enthalpy leaving system} - \text{enthalpy entering system}$$
$$= (m_f h_{fo} + m_w h_{wo}) - (m_f h_{fi} + m_w h_{wi})$$
$$= m_f(h_{fo} - h_{fi}) + m_w(h_{wo} - h_{wi})$$
$$= 80 \frac{\text{lb}}{\text{min}} (110 - 960) \frac{\text{Btu}}{\text{lb}} + 1{,}600 \frac{\text{lb}}{\text{min}} \left(40 \frac{\text{Btu}}{\text{lb}} \right)$$
$$= -68{,}000 \text{ Btu/min} + 64{,}000 \text{ Btu/min}$$
$$= -4{,}000 \text{ Btu/min, i.e. flowing from the system}$$

(b) Consider the condensing fluid alone. Q represents the total heat energy transfer from the fluid. Therefore the fluid alone,

$$Q = m_f(h_{fo} - h_{fi})$$
$$= 80 \frac{\text{lb}}{\text{min}} (100 - 960) \frac{\text{Btu}}{\text{lb}}$$
$$= -68{,}000 \text{ Btu/min}$$

i.e. 68,000 Btu/min leaving the fluid.

Of this, $\left(1{,}600 \dfrac{\text{Btu}}{\text{lb}} \times 40 \dfrac{\text{lb}}{\text{min}} \right) = 64{,}000$ Btu/min pass into the cooling water. Therefore the remainder, i.e. $68{,}000 - 64{,}000 = 4{,}000$ Btu/min pass into the atmosphere.

2.8 Turbine

A turbine is a device which uses a pressure drop to produce work energy which is used to drive an external load.

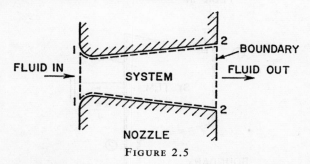

FIGURE 2.5

The steady flow energy equation gives

$$Q - \frac{W}{J} = m\left[(h_2 - h_1) + \frac{c_2{}^2 - c_1{}^2}{2 g_c J} \right]$$

Points to note

(i) The average velocity of flow of fluid through a turbine is normally high, and the fluid passes quickly through the turbine. It may be assumed that, because of this, heat energy does not have time to flow into or out of the fluid during its passage through the turbine, and hence $Q = 0$.

(ii) Although velocities are high the difference between them is not large, and the term representing the change in kinetic energy may be neglected.

(iii) W will be the amount of external work energy produced per second.

The steady flow energy equation becomes

$$-\frac{W}{J} = m(h_2 - h_1)$$

or

$$\frac{W}{J} = m(h_1 - h_2) \qquad (2.6)$$

i.e. positive work since $h_1 > h_2$

Ex. 2.3. A fluid flows through a turbine at the rate of 100 lb/min. Across the turbine the specific enthalpy drop of the fluid is 250 Btu/lb and the turbine loses 2,000 Btu/min in the form of heat energy. Determine the horsepower produced by the turbine, assuming that changes in kinetic and potential energy may be neglected.

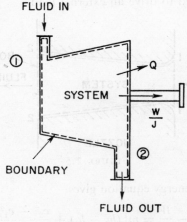

FLUID IN

SYSTEM

BOUNDARY

FLUID OUT

FIGURE E.2.3

Steady flow energy equation gives:

$$Q - \frac{W}{J} = m\left[(h_2 - h_1) + \left(\frac{c_2^2 - c_1^2}{2g_cJ}\right) + \frac{g}{g_c}\left(\frac{z_2 - z_1}{J}\right)\right]$$

Q = heat energy flow into system = $-$ 2,000 Btu/min

W/J = work energy flow from system Btu/min

m = fluid flow rate = 100 lb/min

$h_2 - h_1 = -$ 250 Btu/lb

$(c_2^2 - c_1^2)/2g_cJ$ and $(z_2 - z_1)g/Jg_c$ are negligible.

Therefore the steady flow energy equation becomes

$$- 2,000 - \frac{W}{J} = 100 \frac{lb}{min}\left[- 250 \frac{Btu}{lb}\right]$$

$$\therefore \qquad \frac{W}{J} = (25,000 - 2,000) \text{ Btu/min}$$

$$= 23,000 \text{ Btu/min}$$

$$\therefore \qquad W = 23,000 \frac{Btu}{min} \times 778 \frac{ft\ lbf}{Btu}$$

$$= (23,000 \times 778) \frac{ft\ lbf}{min}$$

$$= \frac{23,000 \times 778}{33,000} \text{ hp}$$

$$= 542 \text{ hp}$$

2.9 Compressor

The action of a compressor is the reverse of that of a turbine, i.e. it uses external work energy to produce a pressure rise. In applying the steady flow energy equation to a compressor, exactly the same arguments are used as for a turbine and the equation becomes

$$- \frac{W}{J} = m(h_2 - h_1) \qquad (2.7)$$

Since $h_2 > h_1$, W will be found to be negative.

Ex. 2.4. A compressor delivers fluid at the rate of 100 lb/min. At the inlet to the compressor the specific enthalpy of the fluid is 20 Btu/lb, and at outlet from the compressor the specific enthalpy of the fluid is 75 Btu/lb. If 100 Btu/min of heat energy are lost to the surroundings by the compressor, determine the horse power required to drive the compressor if the efficiency of the drive is 85%.

$$\text{Flow rate of fluid} = 100 \text{ lb/min}$$

$$= \frac{100}{60} = 1\cdot67 \text{ lb/sec}$$

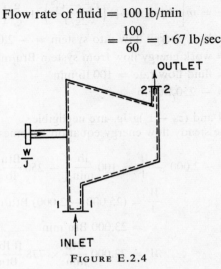

FIGURE E.2.4

Steady flow energy equation is

$$Q - \frac{W}{J} = m\left[(h_2 - h_1) + \frac{c_2{}^2 - c_1{}^2}{2g_cJ}\right]$$

$Q = -100 \text{ Btu/min} = -1\cdot67 \text{ Btu/sec}$
$W/J = \text{work energy flow (Btu/sec)}$
$h_1 = 20 \text{ Btu/lb}$
$h_2 = 75 \text{ Btu/lb}$
$m = 1\cdot67 \text{ lb/sec}$

$\dfrac{c_2{}^2 - c_1{}^2}{2g_cJ}$ may be neglected (see §2.8).

Substituting in the steady flow energy equation gives

$$\left(-1\cdot67 \frac{\text{Btu}}{\text{sec}}\right) - \left(\frac{W}{J}\right) = 1\cdot67 \frac{\text{lb}}{\text{sec}}\left[(75 - 20)\right] \frac{\text{Btu}}{\text{lb}}$$

$$\therefore \qquad \frac{W}{J} = -1\cdot67 - (1\cdot67 \times 55) \text{ Btu/sec}$$

$$= -93\cdot5 \text{ Btu/sec}$$

$$\therefore \qquad W = -93\cdot5 \times 778 \text{ ft lbf/sec}$$

$$= -\frac{93\cdot5 \times 778}{550} \text{ hp}$$

$$= -132\cdot2 \text{ hp}$$

(N.B. The −ve sign indicated work energy is required by the compressor.)

Since the efficiency of the drive is 85%

horsepower required by compressor $= 132 \cdot 2 \times \dfrac{100}{85}$

$$= \underline{155 \cdot 7} \text{ hp}$$

2.10 Nozzle

A nozzle utilises a pressure drop to produce an increase in the kinetic energy of the fluid.

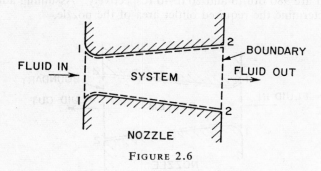

FIGURE 2.6

The steady flow energy equation gives

$$Q - \frac{W}{J} = m\left[(h_2 - h_1) + \frac{c_2{}^2 - c_1{}^2}{2g_cJ}\right]$$

Points to note

  (i) The average velocity of flow through a nozzle is high, hence the fluid spends only a short time in the nozzle. For this reason, it may be assumed that there is insufficient time for heat energy to flow into or out of the fluid during its passage through the nozzle, i.e. $Q = 0$.

 (ii) Since a nozzle has no moving parts, no work energy will be transferred to or from the fluid as it passes through the nozzle, i.e. $W = 0$. Hence equation becomes

$$0 = m\left[(h_2 - h_1) + \frac{c_2{}^2 - c_1{}^2}{2g_cJ}\right]$$

Often c_1 is negligible compared with c_2. In this case the equation becomes

$$0 = m\left[(h_2 - h_1) + \frac{c_2^2}{2g_cJ}\right]$$

or

$$\frac{c^2}{2g_cJ} = h_1 - h_2$$

or

$$c_2 = \sqrt{[2g_cJ(h_1 - h_2)]} \tag{2.8}$$

Ex. 2.5. Fluid with a specific enthalpy of 1,200 Btu/lb enters a horizontal nozzle with negligible velocity at the rate of 30 lb/sec. At the outlet from the nozzle the specific enthalpy and specific volume of the fluid are 980 Btu/lb and 20 ft³/lb respectively. Assuming adiabatic flow, determine the required outlet area of the nozzle.

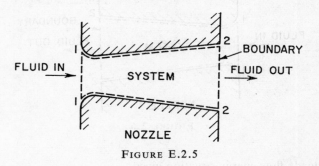

FIGURE E.2.5

Steady flow energy equation gives

$$Q - \frac{W}{J} = m\left[(h_2 - h_1) + \frac{c_2^2 - c_1^2}{2g_cJ}\right]$$

Applied to a nozzle (see §2.9) this becomes

$$0 = m\left[(h_2 - h_1) + \frac{c_2^2 - c_1^2}{2g_cJ}\right]$$

Since the inlet velocity c_1 is negligible, this may be written

$$c_2 = \sqrt{[2g_cJ(h_1 - h_2)]}$$

$$= \sqrt{\left[2 . 32{\cdot}2 \times 778 \times (1,200 - 980)\,\frac{\text{lb}}{\text{lbf}}\frac{\text{ft}}{\text{sec}^2}\frac{\text{ft lbf}}{\text{Btu}}\frac{\text{Btu}}{\text{lb}}\right]}$$

$$= 3,315 \text{ ft/sec}$$

Applying the equation of continuity at outlet gives

$$m = \frac{A_2 c_2}{v_2} \text{ (see § 2.12)}$$

$$\therefore 30\,\frac{\text{lb}}{\text{sec}} = \frac{A_2 \times 3{,}315\,\text{ft/sec}}{20\,\text{ft}^3\,\text{lb}}$$

$$\therefore A = \frac{30 \times 20}{3{,}315}\,\text{ft}^2$$

$$= 0\cdot181\,\text{ft}^2$$

$$= 26\cdot1\,\text{in}^2$$

2.11 Throttling

A throttling process is one in which the fluid is made to flow through a restriction, e.g. a partially opened valve or orifice, causing a considerable drop in the pressure of the fluid.

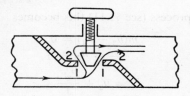

FIGURE 2.7

The steady flow energy equation gives

$$Q - \frac{W}{J} = m\left[(h_2 - h_1) + \frac{c_2{}^2 - c_1{}^2}{2g_c J}\right]$$

Points to note

    (i) Since the throttling takes place over a very small distance, the available area through which heat energy can flow is very small, and it is normally assumed that no energy is lost by heat transfer, i.e. $Q = 0$

   (ii) Since there are no moving parts, no energy can be transferred in the form of work energy, i.e. $W/J = 0$

  (iii) The difference between c_1 and c_2 will not be great and consequently the term representing the change in kinetic energy is normally neglected

The steady flow energy equation becomes

$$0 = m(h_2 - h_1) \text{ or } h_2 = h_1 \qquad (2.9)$$

i.e. during a throttling process the enthalpy remains constant.

Ex. 2.6. A fluid flowing along a pipe line undergoes a throttling process from 150 lbf/in² to 15 lbf/in² in passing through a partially open valve. Before throttling, the specific volume of the fluid is 5 ft³/lb and after throttling is 29·7 ft³/lb. Determine the change in specific internal energy during the throttling process.

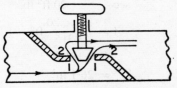

FIGURE E.2.6

Steady flow energy equation

$$Q - \frac{W}{J} = m\left[(h_2 - h_1) + \frac{c_2{}^2 - c_1{}^2}{2g_cJ}\right]$$

For a throttling process (see §2.11) this becomes

$$m(h_2 - h_1) = 0$$
$$\therefore h_2 = h_1$$

But

$$h_2 = u_2 + \frac{P_2 v_2}{J}$$

and

$$h_1 = u_1 + \frac{P_1 v_1}{J}$$

Therefore change in specific internal energy

$$= u_2 - u_1$$
$$= \left(h_2 - \frac{P_2 v_2}{J}\right) - \left(h_1 - \frac{P_1 v_1}{J}\right)$$
$$= (h_2 - h_1) - \left(\frac{P_2 v_2}{J} - \frac{P_1 v_1}{J}\right)$$
$$= 0 - \left(\frac{15 \times 144 \times 29 \cdot 7}{778} - \frac{150 \times 144 \times 5}{778}\right) \frac{\text{lbf}}{\text{ft}^2} \cdot \frac{\text{ft}^3}{\text{lb}} \cdot \frac{\text{Btu}}{\text{ft lbf}}$$
$$= -(82 \cdot 5 - 139) \text{ Btu/lb}$$
$$= +56 \cdot 5 \text{ Btu/lb}$$

2.12 Equation of Continuity

This is an equation which is often used in conjunction with the steady flow energy equation. It is based on the fact that if a system is in a steady state, then the mass of fluid passing any section during a specified time must be constant.

Consider a mass of m lb/sec flowing through a system in which all conditions are steady as illustrated in Fig. 2.8.

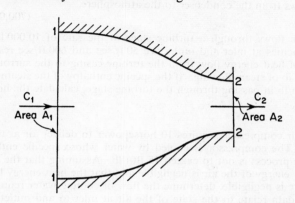

FIGURE 2.8

Let A_1, A_2 represent flow areas in ft^2 at inlet and outlet respectively.

Let v_1, v_2 represent specific volumes in ft^3/lb at inlet and outlet respectively.

Let c_1, c_2 represent velocities in ft/sec, at inlet and outlet respectively.

$$\text{Then mass flowing per sec} = \frac{\text{volume flowing per sec}}{\text{volume per lb}} \frac{\text{ft}^3/\text{sec}}{\text{ft}^3/\text{lb}}$$

$$= \frac{A_1 c_1}{v_1} \text{ lb/sec at inlet}$$

$$= \frac{A_2 c_2}{v_1} \text{ lb/sec at outlet}$$

i.e. $$m = \frac{A_1 c_1}{v_1} = \frac{A_2 c_2}{v_2} \qquad (2.10)$$

EXERCISES ON CHAPTER 2

N.B. All pressures are absolute.

1. A boiler uses coal at the rate of 3 ton/h in producing steam with a specific enthalpy of 1,160 Btu/lb from feed water with a specific enthalpy of 120 Btu/lb. The combustion of 1 lb of coal produces 12,000 Btu, of which 80% is useful in producing steam. Calculate the rate at which steam is produced.

(62,000 lb/h)

2. Fluid with a specific enthalpy of 980 Btu/lb enters a condenser at the rate of 10,000 lb/h, and leaves with a specific enthalpy of 70 Btu/lb. If the enthalpy of the cooling water circulating through the condenser tubes increases at the rate of 140,000 Btu/min, determine the rate at which heat energy flows from the condenser to the atmosphere.

(700,000 Btu/h)

3. Steam flows through a turbine stage at the rate of 10,000 lb/h. The steam velocities at inlet and outlet are 50 ft/sec and 600 ft/sec respectively. The rate of heat energy flow from the turbine casing to the surroundings is 10 Btu per lb of steam flowing. If the specific enthalpy of the steam decreases by 180 Btu/lb in passing through the turbine stage, calculate the horsepower developed.

(640 h.p.)

4. An air compressor requires 10 horsepower to deliver air at the rate of 5 lb/min. The compressor is cooled by water whose specific enthalpy rise during the process is not to exceed 15 Btu/lb. Assuming that the change in the kinetic energy of the air is negligible and that the heat energy loss to the atmosphere is negligible, determine the flow of cooling water required. The following data relate to the state of the air at inlet to and outlet from the compressor:

Section	Pressure	Specific volume	Specific internal energy
	lbf/in^2	ft^3/lb	Btu/lb
Inlet	13·5	14·7	102
Outlet	100	2·96	188

(384 lb/h)

5. A nozzle is supplied with steam having a specific enthalpy of 1,194 Btu/lb at the rate of 20 lb/min. At outlet from the nozzle the velocity of the steam is 3,500 ft/sec. Assuming that the inlet velocity of the steam is negligible and that the process is adiabatic, determine:

    (a) The specific enthalpy of the steam at the nozzle exit
    (b) The outlet area required if the final specific volume of the steam is 300 ft³/lb

(949 Btu/lb; 4·12 in²)

6. Fluid at 150 lbf/in² having a specific volume of 2·9 ft³/lb is throttled to a pressure of 14·7 lbf/in². If the specific volume of the fluid after throttling is 1·72 ft³/lb, calculate the change in specific internal energy during the process.

(+76·7 Btu/lb)

7. Gas with a specific enthalpy of 360 Btu/lb flows through a turbine at the rate of 1,500 lb/min, and leaves with a specific enthalpy of 275 Btu/lb. Assuming that the process is adiabatic and that the kinetic energy terms are negligible, determine the horsepower output of the turbine.

(3,000 hp)

8. The following data refer to a gas nozzle:

Section	Pressure	Specific volume	Specific internal energy
	lbf/in^2	ft^3/lb	Btu/lb
Inlet	70	8·35	312
Intermediate	37·8	13·2	266
Outlet	14·7	28·0	220

It may be assumed that the inlet velocity is negligible, and that the process is adiabatic. If the area of flow at the intermediate section is 1 in², calculate the rate of gas flow through the nozzle and the required outlet area.

(55·6 lb/min; 1·5 in²)

9. A fluid A having a specific enthalpy of 980 Btu/lb flows into a rigid mixing tank at the rate of 100 lb/min. A second fluid B having a specific enthalpy of 70 Btu/lb also flows into the tank at a steady rate. During mixing, the rate of heat energy loss to the atmosphere is 1,200 Btu/min. If the specific enthalpy of the mixture flowing from the tank is 450 Btu/lb, and the fluid level in the tank remains steady, calculate the rate of flow of fluid B.

(136·4 lb/min)

10. A steam turbine uses 8,000 lb of steam per hour. At inlet to the turbine the steam has a velocity of 90 ft/sec, and a specific enthalpy of 1,300 Btu/lb. The steam leaves the turbine with a velocity of 600 ft/sec and a specific enthalpy of 950 Btu/lb. If the process is adiabatic, calculate the output of the turbine.

(1,080 hp)

11. Air enters a centrifugal compressor at 14·5 lbf/in² and leaves at 30 lbf/in². During the passage of the air through the compressor, the specific internal energy of the air increases by 24 Btu/lb, and the specific volume decreases from 13·2 ft³/lb to 8·1 ft³/lb. Assuming that the flow is adiabatic and that the changes in kinetic energy are negligible, calculate the horsepower required to drive the compressor when the air flow rate is 300 lb/min.

(238 hp)

3: Non-flow Processes

In thermodynamics, instances often arise where the application of the general energy equation may be simplified. Once a fluid has entered a system, it may be possible for it to undergo a series of processes in which the fluid does not flow. An example of this is the cylinder of an internal combustion engine. On the suction stroke, the working fluid

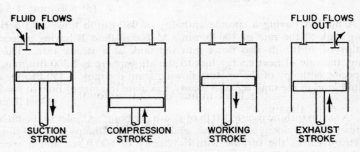

FIGURE 3.1

flows into the cylinder, which is then temporarily sealed. Whilst the cylinder is sealed, the fluid is compressed by the piston moving into the cylinder, after which heat energy is supplied so that the fluid possesses sufficient energy to force the piston back down the cylinder, causing the engine to do external work. The exhaust valve is then opened and the fluid is made to flow out of the cylinder into the surroundings. Processes which are undergone by a system when the working fluid cannot cross the boundary are called non-flow processes. These processes occur during the compression and the working stroke in the above example. The simplified general energy equation which may be applied to these processes is called the Non-flow Energy Equation.

3.1 Non-flow Energy Equation

In the preceding chapter, one form of the general energy equation was shown to be, for a mass flow of m lb/sec,

$$m\left(\frac{z_1}{J} + \frac{c_1^2}{2gJ} + u_1 + \frac{P_1v_1}{J}\right) + Q - \frac{W}{J} = m\left(\frac{z_2}{J} + \frac{c_2^2}{2gJ} + u_2 + \frac{P_2v_2}{J}\right)$$

34

If the fluid is undergoing a non-flow process from state (1) to state (2) then the terms P_1v_1/J and P_2v_2/J (which represented the amount of work energy required to introduce and expel the fluid from the system) will be zero, since the fluid will already be in the system, and will still be in the system at the end of the process. For the same reason, the changes in kinetic and potential energies of the fluid will also be zero. Thus the equation becomes

$$U_1 + Q - \frac{W}{J} = U_2$$

where $U_1 = mU_1$ and $U_2 = mU_2$, or

$$U_2 - U_1 = Q - \frac{W}{J} \qquad (3.1)$$

In words, this equation states that in a non-flow process, the change in the internal energy of the fluid is equal to the nett amount of heat energy supplied to the fluid minus the nett amount of work energy flowing from the fluid.

This equation is known as the non-flow energy equation, and it will now be shown how this may be applied to various non-flow processes.

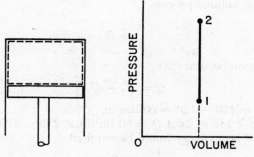

FIGURE 3.2

3.2 Constant Volume Process

Consider the amount of fluid contained within the boundary. If the volume is to remain constant, then the piston will not be able to move. That is, no energy transfer in the form of work energy can take place, i.e. $W/J = 0$. Therefore, if there are to be any energy changes, they must occur by the fluid either gaining or losing energy by the transfer of heat energy. If U_1 represents the initial energy of the fluid, U_2 represents the final internal energy of the fluid, and Q represents the nett amount of heat energy passing into the fluid, then applying the non-flow energy equation

$$Q - \frac{W}{J} = U_2 - U_1$$

gives

$$Q - 0 = U_2 - U_1$$

i.e.

$$Q = U_2 - U_1 \qquad (3.2)$$

for a constant volume process.

This result, which is important and should be remembered, shows that the net amount of heat energy supplied to or taken from a fluid during a constant volume process is equal to the change in the internal energy of the fluid.

Ex. 3.1. The specific internal energy of a fluid is increased from 120 Btu/lb to 180 Btu/lb during a constant volume process. Determine the amount of heat energy required to bring about this increase for 2 lb of fluid.

The non-flow energy equation is

$$Q - \frac{W}{J} = U_2 - U_1$$

For a constant volume process

$$\frac{W}{J} = 0$$

and the equation becomes

$$Q = U_2 - U_1$$

In Btu/lb, $Q = 180 - 120 = 60$ Btu/lb.

Therefore for 2 lb of fluid $Q = 60$ Btu/lb $\times$ 2 lb $= 120$ Btu.
i.e. 120 Btu of heat energy would be required.

3.3 Constant Pressure Process

Consider the quantity of fluid contained within the boundary at state (1).

If the pressure on the fluid remains at a constant pressure P lbf/ft², then as was shown in Chapter 1, the amount of energy transferred from the fluid in the form of work energy in moving the piston reversibly from volume V_1 ft³ to volume V_2 ft³ is given, in heat units, by

$$\frac{W}{J} = \frac{P(V_2 - V_1)}{J}$$

If U_1 and U_2 represent the initial and final internal energy of the fluid respectively, and Q represents the amount of heat energy passing into the fluid, then applying the non-flow energy equation

$$Q - \frac{W}{J} = U_2 - U_1$$

gives

$$Q - \frac{P(V_2 - V_1)}{J} = U_2 - U_1 \qquad (3.3)$$

∴

$$Q = U_2 - U_1 + \frac{P(V_2 - V_1)}{J}$$

$$= \left(U_2 + \frac{P_2 V_2}{J}\right) - \left(U_1 + \frac{P_1 V_1}{J}\right)$$

since

$$P = P_1 = P_2$$

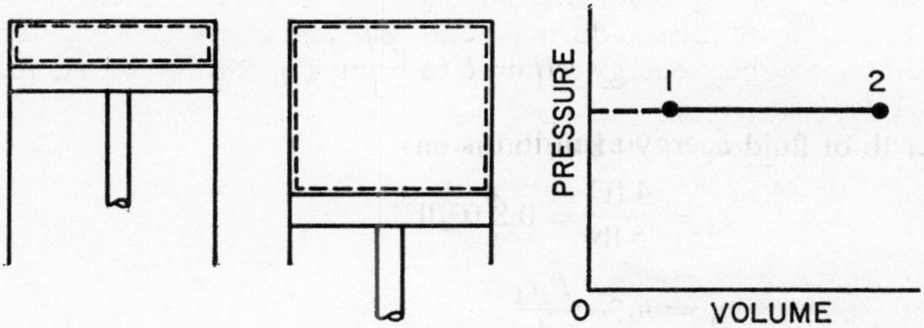

FIGURE 3.3

As stated in Chapter 2, terms of the form $[U + (PV/J)]$ are given the name enthalpy and denoted by H or mh, where m is the mass of fluid, and h is the specific enthalpy.

Therefore

$$H_1 = U_1 + \frac{P_1 V_1}{J}$$

and

$$H_2 = U_2 + \frac{P_2 V_2}{J}$$

i.e.

$$Q = H_2 - H_1 = m(h_2 - h_1) \qquad (3.4)$$

for a reversible constant pressure process.

This result is also important and should be remembered. It shows that the nett amount of heat energy supplied to or taken from a fluid during a reversible constant pressure process is equal to the change of enthalpy of the fluid during that process.

Ex. 3.2. Five pounds of a fluid having a volume of 4 ft³ are contained in a cylinder at a constant pressure of 100 lbf/in². Heat energy is supplied to the fluid until the volume becomes 8 ft³. If the initial and final specific enthalpies of the fluid are 90 Btu/lb and 120 Btu/lb respectively, determine, (a) the quantity of heat energy supplied to the fluid, (b) the change in internal energy of the fluid.

Referring to Fig. 3.3, $P = 100 \text{ lbf/in}^2$, $V_1 = 4 \text{ ft}^3$ and $V_2 = 8 \text{ ft}^3$. For a constant pressure process, from equation (3.4)

heat energy supplied = change in enthalpy of fluid

i.e. $$Q = H_2 - H_1 = m(h_2 - h_1)$$

$$= 5 \text{ lb}\left(120 \frac{\text{Btu}}{\text{lb}} - 90 \frac{\text{Btu}}{\text{lb}}\right) = 150 \frac{\text{Btu}}{\text{lb}}$$

Part (b) of the question may be solved by either of two methods.

Method 1

$$H = U + \frac{PV}{J}$$

Per lb of fluid $h_1 = 90$ Btu/lb (given)

$$v_1 = \frac{4 \text{ ft}^3}{5 \text{ lb}} = 0.8 \text{ ft}^3/\text{lb}$$

$\therefore$ $$u_1 = h_1 - \frac{P_1 v_1}{J}$$

$$= 90 \frac{\text{Btu}}{\text{lb}} \text{lb} - \frac{(100 \times 144)(\text{lbf/ft}^2) \times 0.8 \text{ (ft}^3/\text{lb)}}{778 \text{ ft } 1\text{bf/Btu}}$$

$$= 90 - 14.8 = 75.2 \text{ Btu/lb}$$

Similarly $h_2 = 120$ Btu/lb (given)

$$v_2 = \frac{8 \text{ ft}^3}{5 \text{ lb}} = 1.6 \text{ ft}^3/\text{lb}$$

$\therefore$ $$u_2 = h_2 - \frac{P_2 v_2}{J}$$

$$= 120 \frac{\text{Btu}}{\text{lb}} - \frac{100 \times 144 \times 1.6}{778} \frac{\text{Btu}}{\text{lb}}$$

$$= 120 - 29.6 = 90.4 \text{ Btu/lb}$$

$\therefore$ change in internal energy per lb = $u_2 - u_1$
$$= 90.4 - 75.2 = 15.2 \text{ Btu/lb}$$

$\therefore$ change in internal energy for 5 lb = $15.2 \dfrac{\text{Btu}}{\text{lb}} \times 5 \text{ lb}$

$$= 76 \text{ Btu}$$

Method 2

For a constant pressure process

$$\frac{W}{J} = \frac{P(V_2 - V_1)}{J}$$

$$\therefore \quad \frac{W}{J} = \frac{(100 \times 144)(\text{lbf/ft}^2) \times (8 \text{ ft}^3 - 4 \text{ ft}^3)}{778 \text{ ft lbf/Btu}}$$

$$= \frac{100 \times 144 \times 4}{778} \frac{\text{lbf}}{\text{ft}^2} \times \text{ft}^3 \times \frac{\text{Btu}}{\text{ft lbf}}$$

$$= 74 \text{ Btu}$$

Applying the non-flow energy equation

$$Q - \frac{W}{J} = U_2 - U_1$$

gives $U_2 - U_1 = 150 \text{ Btu} - 74 \text{ Btu} = 76 \text{ Btu}$

3.4 Adiabatic Process

In an adiabatic process, no heat energy is allowed to cross the boundary (usually by insulating the boundary). Q is, therefore, zero. Let U_1 and $U_2 =$ initial and final internal energies of the system respectively, and let $W/J =$ net amount of work energy passing from the system. Applying the non-flow energy equation gives

$$0 - \frac{W}{J} = U_2 - U_1$$

i.e.
$$\frac{W}{J} = U_1 - U_2 \tag{3.5}$$

This means that if W/J is positive then $(U_1 - U_2)$ must also be positive, i.e. the final internal energy will be less than the initial internal energy. Therefore, if a system does external work on the surroundings during a non-flow adiabatic process, the work is done at the expense of the internal energy of the system. Conversely, if work is done on the system during a non-flow adiabatic process, then the internal energy of the system will be increased.

Ex. 3.3. During an adiabatic expansion, 0·5 lb of fluid transfers 32,000 ft lbf of work energy. Determine the change in the specific internal energy of the fluid during the process.

The non-flow energy equation gives

$$Q - \frac{W}{J} = U_2 - U_1$$

For an adiabatic process this becomes

$$0 - \frac{W}{J} = U_2 - U_1$$

$$\therefore U_2 - U_1 = \frac{32{,}000 \text{ ft lbf}}{778 \text{ ft lbf/Btu}} = -41 \cdot 2 \text{ Btu}$$

i.e. change of internal energy for 0·5 lb = − 41·2 Btu

$$\therefore \text{ change of internal energy per lb} = -\frac{41 \cdot 2 \text{ Btu}}{0 \cdot 5 \text{ lb}} = -82 \cdot 4 \text{ Btu/lb}$$

i.e. the internal energy decreases during the process.

3.5 Polytropic Process

This is the most general type of process, in which both heat energy and work energy cross the boundary of the system. It is represented by an equation of the form

$$PV^n = \text{constant}$$

where n, known as the index of expansion (or compression) is a constant. If a fluid changes from an initial state (1) to a final state (2) by such a process, the amount of work energy crossing the boundary is given by

$$\frac{W}{J} = \frac{P_1 V_1 - P_2 V_2}{J(n-1)} \tag{3.6}$$

(for proof see Appendix 2.1).

Therefore, applying the non-flow energy equation to a polytropic process gives

$$Q - \frac{W}{J} = U_2 - U_1$$

i.e. $$Q - \frac{P_1 V_1 - P_2 V_2}{J(n-1)} = U_2 - U_1 \tag{3.7}$$

Ex. 3.4. A cylinder contains 0·15 lb of fluid having a pressure of 15 lbf/in², a volume of 2 ft³ and a specific internal energy of 88 Btu/lb. After polytropic compression, the pressure and volume of the fluid are 135 lbf/in² and 0·37 ft³ respectively, and the specific internal energy is 160 Btu/lb.

Determine, (a) the amount of work energy required for the compression, (b) the quantity and direction of the heat energy flowing during the compression.

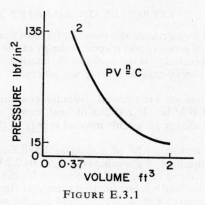

FIGURE E.3.1

(a) For a polytropic process, $PV^n = $ constant, i.e. $P_1V_1^n = P_2V_2^n$. In the given case

$$(15 \times 144) \times 2^n = (135 \times 144) \times (0.37)^n$$

$$\therefore \qquad \frac{2^n}{0.37^n} = \frac{135 \times 144}{15 \times 144}$$

i.e. $\qquad\qquad 5.4^n = 9$

$$\therefore \qquad\qquad n = 1.302$$

$$\frac{W}{J} = \frac{P_1V_1 - P_2V_2}{J(n-1)}$$

$$= \frac{(15 \times 144)(\mathrm{lbf/ft^2}) \times 2\ \mathrm{ft^3} - (135 \times 144)(\mathrm{lbf/ft^2}) \times 0.37\ \mathrm{ft^3}}{(1.302 - 1) \times 778\ (\mathrm{ft\ lbf/Btu})}$$

$$= -12.2\ \mathrm{Btu}$$

The negative sign indicates that work energy would flow into the system during the process.

(b) Non-flow energy equation gives

$$Q - \frac{W}{J} = U_2 - U_1$$

$$Q - (-12.2)\ \mathrm{Btu} = \left(160\ \frac{\mathrm{Btu}}{\mathrm{lb}} \times 0.15\ \mathrm{lb}\right) - \left(88\ \frac{\mathrm{Btu}}{\mathrm{lb}} \times 0.15\ \mathrm{lb}\right)$$

i.e. $\qquad Q + 12.2\ \mathrm{Btu} = 24\ \mathrm{Btu} - 13.2\ \mathrm{Btu} = 10.8\ \mathrm{Btu}$

$$\therefore \qquad\qquad Q = -1.4\ \mathrm{Btu}$$

The negative sign indicates that heat energy will flow out of the fluid during the process.

EXERCISES ON CHAPTER 3

1. During a non-flow process, the volume of 3 lb of fluid remains constant at 9·5 ft³. The initial pressure and specific enthalpy of the fluid are 100 lbf/in² and 200 Btu/lb respectively. If the final specific internal energy is 110 Btu/lb, calculate the heat energy transfer during the process.

(−93·9 Btu)

2. A sealed sphere of 3 ft internal diameter contains a fluid having a specific volume of 6 ft³/lb. If 100 Btu of heat energy are supplied to the fluid, calculate the change in specific internal energy of the fluid.

(+42·2 Btu/lb)

3. A cylinder contains 0·5 ft³ of a fluid at 100 lbf/in² having a specific enthalpy of 298·4 Btu/lb, and a specific volume of 0·02 ft³/lb. Heat energy is supplied until the volume of the fluid becomes 10 ft³, the pressure remaining at 100 lbf/in². If the final specific internal energy of the fluid is 380 Btu/lb, calculate: (a) the mass of fluid, (b) the final specific volume, (c) the heat energy supplied, (d) the work energy transferred.

(25 lb; 0·4 ft³/lb; 2,225 Btu; +176 Btu)

4. During a non-flow process, 1 lb of steam is cooled at constant pressure from a volume of 10·5 ft³ to a volume of 1 ft³. If the energy transfers during the process are 846 Btu in the form of heat energy and 54,700 ft lbf in the form of work energy, determine the pressure at which the process takes place and, also, the change in specific internal energy.

(40 lbf/in²; −775·7 Btu/lb)

5. If a system undergoes a non-flow adiabatic process during which 66,000 ft lbf of work energy are transferred from the surroundings to the system and the internal energy of the system changes by 16 Btu/lb, calculate the mass of fluid contained in the system.

(5·3 lb)

6. A fluid contained in a cylinder undergoes an adiabatic process in which 30,000 ft lbf of work energy are transferred to the surroundings, followed by a constant volume process in which 50 Btu of heat energy are transferred to the surroundings. If the mass of fluid in the cylinder is 2 lb, calculate the change in specific internal energy during each process, and also the overall change in internal energy of the fluid.

(−19·3 Btu/lb; −25 Btu/lb; −88·6 Btu)

7. A quantity of fluid occupies 20 ft³ at 14·7 lbf/in². After a compression which follows the law PV^n = constant, the pressure and volume are 90 lbf/in² and 5 ft³ respectively. Calculate the value of the index n and the work energy transfer. If the internal energy of the fluid increases by 80 Btu during the process, calculate the heat energy transfer.

(1·308; −73,000 ft lbf; −13·8 Btu)

8. A cylinder contains 1·5 lb of fluid at 100 lbf/in². The fluid expands according to the law $PV^{1·27}$ = constant until the pressure is 20 lbf/in². Given that the initial specific volume of the fluid is 4 ft³/lb, calculate the final volume of the fluid. If 30 Btu of heat energy are transferred to the fluid during the expansion, calculate the change in specific internal energy of the fluid.

(21·3 ft³; −59·5 Btu/lb)

4: Properties of Liquids and Vapours

Matter may exist in the form of a solid, liquid, or a gas. A vapour is a gas existing at a temperature at which it can be liquefied by an increase of pressure. Most practical problems in applied thermodynamics are not concerned with the solid phase, but rather with the liquid and gaseous phases. The most common liquid is water with its associated vapour, steam, but other liquids are often used, for example, in refrigerators and heat pumps. Before considering the uses of these liquids and their vapours, a knowledge of their properties and behaviour under various conditions is required.

The properties most commonly used in applied thermodynamics are pressure, temperature, volume, internal energy, enthalpy, and entropy. (For a detailed discussion of entropy, see chapter 6.) Pressure and temperature do not depend upon the mass of the fluid present and are termed *intensive properties*. Volume, internal energy, enthalpy and entropy are dependent on the mass of the fluid present and are termed *extensive properties*. It is usual to consider a unit mass of 1 lb of the liquid and to designate the extensive properties as specific, i.e. specific volume, specific enthalpy, and specific internal energy. If two independent properties are known, then the other properties are also known. Hence, the state of a fluid may be described by plotting two properties as abscissae and ordinates on a graph.

4.1 Constant Pressure Process for a Liquid and its Vapour

Consider 1 lb of water at the freezing point $32°F$ and atmospheric pressure 14.7 lbf/in^2, contained in a cylinder as shown in Fig. 4.1. Let the water be heated up slowly and the changes in volume and temperature be recorded. A plot of these will be as shown in Fig. 4.2. The initial volume at point 1 will be found to be 0.01602 ft^3. As the temperature is increased, the volume will decrease to point 2 at about $39°F$, where the water is at its maximum density. Further increase in temperature causes the volume to increase to 0.01671 at $212°F$ (point 3), the change of volume of the water over this range being very small (approximately 4%). At point 3, the liquid is said to be saturated since any further addition of heat energy at this point will cause the water to boil and start to change into a vapour—a phase change during which the temperature remains constant and the fluid is known as a wet mixture. This constant temperature (i.e. the temperature of the mixture of water and steam between point 3 and point 4) is known as the saturation temperature. At point 4, all the water has just been changed into steam and is described as being a saturated vapour. The volume of the steam

FIGURE 4.1

at point 4 will be found to be 26·80 ft³, a volume which is large compared with the volume of the water at point 3. Further addition of heat energy at point 4 will cause the temperature to increase above the saturation temperature. The vapour is then said to be superheated, and the difference between the actual temperature of the vapour and the

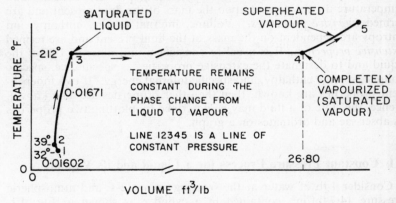

FIGURE 4.2

saturation temperature is known as the degree of superheat. If the foregoing procedure were repeated at a number of higher pressures, and curves of temperature plotted against volume drawn for each pressure on the graph, the resulting family of curves would be as shown in Fig. 4.3.

It should be noticed that as the pressure is increased, the saturation temperature (i.e. the temperature at which boiling occurs) also increases. The line joining all the saturated liquid points 3, 3′, 3″, etc., is known as the saturated liquid line. Similarly, the line joining all the points representing saturated vapour 4, 4′, 4″, etc., is known as the saturated vapour line. The point X where the saturated liquid line and saturated vapour line merge is known as the critical point. At the critical point

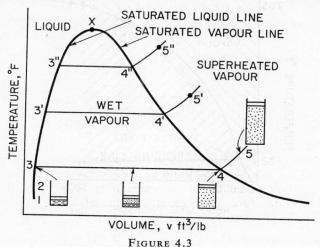

FIGURE 4.3

for water the pressure is 3,206 lbf/in² absolute, the temperature is 705°F, and the volume of the liquid is equal to the volume of the vapour (= 0·0503 ft³/lb).

For pressures higher than the critical pressure the phenomenon of boiling does not exist and the liquid changes immediately into a vapour without the familiar process of boiling.

4.2 Saturation Temperature

The saturation temperature is that at which boiling occurs. At this temperature bubbles of vapour form in the liquid and break through the surface to occupy the space above it as a vapour. Supply of heat at this temperature causes further generation of vapour but does not increase the temperature until all the liquid has been converted into a vapour. Another definition of saturation temperature is that it is the temperature at which the two phases, liquid and vapour can exist in equilibrium with each other. As the pressure is increased so is the saturation temperature, until the critical point is reached. The relationship between the saturation temperature and pressure is shown in Fig. 4.4. The freezing point is depressed slightly as the pressure is increased, falling almost 3 degrees at the critical pressure. The triple point is that at which the three phases, solid, liquid and vapour, can coexist. Below the triple point it is possible for a solid to change into a vapour without first changing into a liquid. This process is known as *sublimation*.

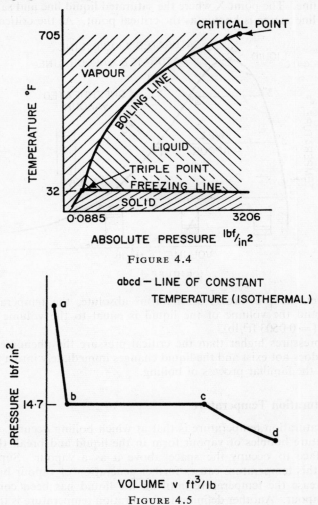

FIGURE 4.4

FIGURE 4.5

4.3 Isothermals on the Pressure–Volume Diagram

Consider 1 lb of water at 212°F contained in a cylinder under a high pressure. Let the pressure be reduced slowly whilst the temperature is maintained at 212°F, and corresponding values of pressure and volume be recorded. A plot of these would be as shown in Fig. 4.5. The line

abcd represents a process in which the temperature remains constant.
Such processes are known as isothermals.

From *a* to *b* the increase in volume will be small for a large decrease
in pressure, since water is almost incompressible. When the pressure
reaches 14·7 lbf/in² at *b*, i.e. the saturation pressure corresponding to
the temperature, the water will start to boil, resulting in a large change
of volume from *b* to *c* whilst the pressure remains constant. As before,
b is known as a saturated liquid point, and *c* is known as a saturated
vapour point (i.e. the point at which all the liquid has just changed into
vapour). Further reduction of the pressure below the value at *b* and *c*

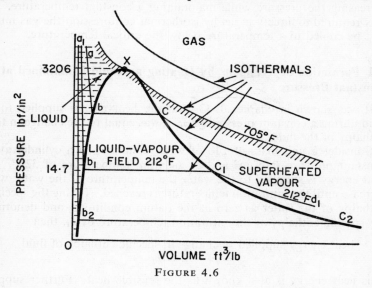

FIGURE 4.6

will result in an increase in volume, the vapour now being superheated.
If the foregoing procedure was repeated for other values of the constant
temperature, and the resulting isothermals plotted on a single *p–v*
diagram, Fig. 4.6 would result.

As in Fig. 4.3 the line joining all saturated liquid states *b*, b_1, b_2, etc.,
is known as the saturated liquid line, and the line joining all saturated
vapour states *c*, c_1, c_2, etc., is known as the saturated vapour line. The
critical point is again denoted by *X*. To the left of the saturated liquid
line and the isothermal passing through the critical point is the compressed liquid phase, so called since the pressure of the liquid at any
point in this region is always greater than the saturation pressure
corresponding to the temperature of the liquid. The region between
the saturated liquid line and the saturated vapour line represents the
liquid–vapour phase. If the point representing the state of the fluid

lies within this region the fluid will consist of a mixture of saturated liquid and saturated vapour and is known as a wet vapour. The region to the right of the saturated vapour line and below the critical isothermal represents the superheated vapour phase, where the temperature of the vapour is higher than the saturation temperature corresponding to the pressure. To the right of the critical isothermal is the region representing the gaseous phase. It should be noted that if the point representing the state of the fluid lies within this region, then the fluid cannot be liquefied by isothermal compression. This fact is sometimes used as an arbitrary definition of a gas, i.e. a gas can never be liquefied by increasing the pressure while maintaining a constant temperature. If it is required to liquefy a gas by isothermal compression the gas must first be cooled to a temperature below the critical temperature.

4.4 Formation of a Vapour by Heating a Liquid Maintained at a Constant Pressure

It was shown in chapter 3, §3, that the heat energy supplied to a fluid during a constant pressure process was equal to the change in the enthalpy of the fluid.

Consider a unit mass of 1 lb of water contained in a cylinder at a constant pressure P lbf/in² and at a datum temperature of 32°F. If heat energy is supplied to the water the temperature of the water will increase until the saturation temperature is reached. Taking the specific enthalpy of the water as zero at the datum conditions, and denoting the specific enthalpy at the saturation temperature by h_f, then

$$\text{heat energy supplied} = \text{change in specific enthalpy of fluid}$$
$$= h_f - 0 = h_f$$

This heat energy is also known as the sensible heat. Further supply of heat energy causes a change of phase at constant temperature, until eventually the liquid is completely transformed into a saturated vapour. If the specific enthalpy of the saturated vapour is denoted by h_g then heat energy supplied during evaporation = change of enthalpy during evaporation = $h_g - h_f$.

This heat energy is sometimes known as the latent heat and is often denoted by L or h_{fg}

i.e. $h_{fg} = h_g - h_f$

or $h_g = h_f + h_{fg}$

If more heat energy is supplied, the vapour becomes superheated and the temperature rises. As before, this additional heat energy will be equal to the change in enthalpy produced and is also equal to mean specific heat of the vapour multiplied by temperature rise.

A plot of temperature against specific enthalpy for various constant pressures for a liquid and its vapour will be as shown in Fig. 4.7.

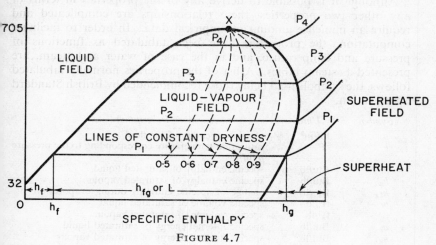

FIGURE 4.7

4.5 Quality of a Vapour

During the change of phase from liquid to vapour, the liquid exists in equilibrium with the vapour, and it can be seen from Fig. 4.6 that the pressure and temperature alone are not sufficient to define completely the state of the mixture, since all state points between b and c have the same pressure and temperature. To determine the exact position of a particular state point in this region, the term quality, or dryness fraction is used. This is defined as the mass of saturated vapour present in unit mass of the mixture, and is generally denoted by x. Thus, for a saturated liquid $x = 0$, for a saturated vapour $x = 1$, and for any condition between these two states, the value of x will lie between 0 and 1.

4.6 Internal Energy of a Liquid or Vapour

In chapter 2, the enthalpy h of a fluid was given by the equation

$$h = u + \frac{Pv}{J}$$

Hence the internal energy of a liquid or vapour may be stated as

$$u = h - \frac{Pv}{J}$$

4.7 Tabulated properties

Although it is possible to derive any of the properties in terms of any other two properties, these relationships are complicated and require an immense amount of empirical data. In order to facilitate computation, the properties have been tabulated as functions of pressure and temperature, and in the case of water and steam, are presented as steam tables. A list of the properties normally tabulated follows, the symbols used being those recommended by British Standard Specification.

Symbol	Units	Description
p	lbf/in²	absolute pressure of the fluid
t	°F	saturation temperature corresponding to the pressure p lbf/in²
h_f	Btu/lb	specific enthalpy of saturated liquid
h_g	Btu/lb	specific enthalpy of saturated vapour
h_{fg}	Btu/lb	change of specific enthalpy during evaporation
v_f	ft³/lb	specific volume of saturated liquid
v_g	ft³/lb	specific volume of saturated vapour
u_f	Btu/lb	specific internal energy of saturated liquid
u_g	Btu/lb	specific internal energy of saturated vapour

Values of the specific entropy of saturated liquid (s_f) and saturated vapour (s_g) are also tabulated. The use and value of these properties will be shown later (see chapter 6).

The datum for the steam tables is taken at temperature of 32°F and the corresponding pressure of 0·0885 lbf/in². At this point the specific volume of saturated liquid is 0·01602 ft³/lb and the specific enthalpy of the saturated liquid is taken as zero. Therefore, at this point

$$u_f = h_f - \frac{Pv_f}{J}$$

$$= 0 - \frac{0.0885 \times 144(\text{lbf/ft}^2) \times 0.01602 \ (\text{ft}^3/\text{lb})}{778 \ \text{ft lbf/Btu}}$$

$$= -0.00026 \ \text{Btu/lb}$$

For most practical purposes this is negligibly small, and the value of u_f at datum is generally taken as zero.

4.8 Properties of a Compressed Liquid

The properties of a compressed liquid can be found from the tables of properties at any pressure and temperature. It is found, however, that except at very high pressures the properties only vary slightly with pressure. For normal pressures, it is sufficiently accurate to take the properties of a compressed liquid as being equal to those of saturated liquid at the same temperature.

4.9 Properties of a Wet Mixture

The specific properties of a wet mixture are obtained using the dryness fraction. If 1 lb of wet mixture has a dryness fraction x, it will consist of x lb of saturated vapour and therefore $(1 - x)$ lb of saturated liquid. The specific enthalpy of the wet mixture h_x will be equal to the enthalpy of x lb of saturated vapour plus the enthalpy of $(1 - x)$ lb of saturated liquid.

i.e.
$$h_x = xh_g + (1 - x)h_f$$
$$xh_g + h_f - xh_f$$
$$= h_f + x(h_g - h_f)$$
$$= h_f + xh_{fg}$$

Similarly
$$u_x = xu_g + (1 - x)u_f = u_f + x(u_g - u_f)$$

and
$$v_x = xv_g + (1 - x)v_f$$

Ex. 4.1. For steam at 150 lbf/in² with a dryness fraction of 0·85, calculate the specific enthalpy, specific volume, and specific internal energy.

Extract from tables

p	t	v_g	u_f	u_g	h_f	h_{fg}	h_g
150	358·4	3·015	330·0	1,110·5	330·5	863·6	1,194·1

(a)
$$h_x = (1 - x)h_f + xh_g$$
$$= (1 - 0·85) \times 330·5 + (0·85 \times 1,194·1)$$
$$= 49·57 + 1,015$$
$$= 1,064·6 \text{ Btu/lb}$$

Alternatively

$$h_x = (1 - x)h_f + xh_g$$
$$= h_f + x(h_g - h_f)$$
$$= h_f + xh_{fg}$$
$$= 330·5 + (0·85 \times 863·6)$$
$$= 330·5 + 734·06$$
$$= 1,064·6 \text{ Btu/lb}$$

(b)
$$v_x = (1 - x)v_f + xv_g$$
$$= (1 - 0·85) \times 0·016 + (0·85 \times 3·015)$$
$$= 0·0024 + 2·563$$
$$= 2·565 \text{ ft}^3\text{/lb}$$

(c)
$$u_x = (1 - x)u_f + xu_g$$
$$= (1 - 0·85)330 + (0·85 \times 1,110·5)$$
$$= 49·5 + 943·9$$
$$= 993·4 \text{ Btu/lb}$$

Ex. 4.2. A cylinder contains 4 ft^3 of steam at 180 lbf/in^2. If the mass of the steam is 2 lb, calculate its dryness fraction, enthalpy, and internal energy.

Extract from tables

p	t	v_g	u_f	u_g	h_f	h_{fg}	h_g
180	373·1	2·532	345·4	1,112·5	346·0	850·8	1,196·8

(a) Volume of 1 lb saturated steam at 180 lbf/in^2 = 2·532 ft^3 = v_g

Volume of 1 lb of steam at given conditions = $\dfrac{4}{2}$ = 2 ft^3 = v_x

Since $v_x = (1 - x)v_f + xv_g$
                 $2 = (1 - x) \times 0·016 + x \times 2·532$

∴ $x = \dfrac{1·984}{2·516} = 0·79$

(b) Specific enthalpy of steam at 180 lbf/in^2, 0·79 dry is given by

$$h_x = (1 - x)h_f + xh_g$$
$$= (1 - 0·79) \times 346·0 + 0·79 \times 1,196·8$$
$$= 72·7 + 945$$
$$= 1,017·7 \text{ Btu/lb}$$

∴ Enthalpy of 2 lb steam at 180 lbf/in^2, 0·79 dry

$$= 1,017·7 \text{ Btu/lb} \times 2 \text{ lb} = 2,035·4 \text{ Btu}$$

(c) Specific internal energy of steam at 180 lbf/in^2, 0·79 dry is given by

$$u_x = (1 - x)u_f + xu_g$$
$$= (1 - 0·79) \times 345·4 + (0·79 \times 1,112·5)$$
$$= 72·5 + 879$$
$$= 951·5 \text{ Btu/lb}$$

∴ Internal energy of 2 lb steam at 180 lbf/in^2, 0·79 dry

$$= 951·5 \text{ (Btu/lb)} \times 2 \text{ lb} = 1,903 \text{ Btu}$$

4.10 Properties of a Superheated Vapour

The values of the specific properties of a superheated vapour are normally listed in separate tables for selected values of pressure and temperature.

The specific volume of a superheated vapour is often omitted from abridged tables of properties. It may then be calculated from the empirical relationship,

$$v = \frac{1·247(h - 835)}{p}$$

where h = specific enthalpy of the superheated steam in Btu/lb, p = pressure in lbf/in^2 and v = specific volume in ft^3/lb.

Ex. 4.3. Calculate the degree of superheat and specific internal energy of steam at 400 lbf/in^2 and 600°F.

Extract from tables

p	t_{sat}	t	v	h
400	444·6	600·	1·4770	1,306·9

Actual temperature of steam = 600°F
Saturation temperature at 400 lbf/in^2 = 444·6°F
∴ Degree of superheat = 600 − 444·6 = 155·4 degF
Specific enthalpy at 400 lbf/in^2, 600°F = h = 1,306·9 Btu/lb
Specific volume at 400 lbf/in^2, 600°F = v = 1·4770 ft^3/lb

∴ Specific internal energy = $u = h - \dfrac{pv}{J}$

$$= 1{,}306{\cdot}9 - \frac{(400 \times 144) \times 1{\cdot}477}{778}$$
$$= 1{,}306{\cdot}9 - 109{\cdot}3$$
$$= 1{,}197{\cdot}6 \text{ Btu/lb}$$

4.11 Use of Interpolation in Tables of Properties

Tables of both saturation properties and superheat properties give the values of properties at selected pressures and temperatures. To obtain values of properties at conditions other than those tabulated, interpolation is used. Values of the properties immediately above and immediately below the required conditions are selected from the tables and the values at the required conditions then obtained by proportioning these tabulated values. Examples 4.4 and 4.5 illustrate the process of interpolation.

Ex. 4.4. Steam at 200 lbf/in^2 has a specific volume of 3 ft^3/lb. Calculate the temperature, specific enthalpy, and specific internal energy.

Extract from tables

p (t_{sat})		t	
		500°F	600°F
200	v	2·726	3·060
(381·8)	h	1,268·9	1,322·1

From the above, 100 degF increase of temperature produces an increase of specific volume of 3·060 − 2·726 = 0·334 ft^3/lb.

Specific volume of given steam = 3 ft³/lb

i.e. $3 - 2 \cdot 726 = 0 \cdot 274$ ft³/lb increase above the specific volume at 500°F.

$$\therefore \text{ increase in temperature above } 500°\text{F} = \frac{0 \cdot 274}{0 \cdot 334} \times 100 = 82 \text{ degF}$$

$\therefore$ temperature of steam = $500 + 82 = 582°$F

Similarly, an increase of temperature from 500°F to 600°F produces an increase in specific enthalpy of $1,322 \cdot 1 - 1,268 \cdot 9 = 53 \cdot 2$ Btu/lb

$\therefore$ an increase of 82 degF increases specific enthalpy by $53 \cdot 2 \times \dfrac{82}{100}$

$= 43 \cdot 7$ Btu/lb

$\therefore$ specific enthalpy of steam = $1,268 \cdot 9 + 43 \cdot 7 = 1,312 \cdot 6$ Btu/lb

$$\text{Specific internal energy} = h - \frac{pv}{J}$$

$$= 1,312 \cdot 6 - \frac{(200 \times 144) \times 3}{778}$$

$$= 1,312 \cdot 6 - 111$$

$$= 1,201 \cdot 6 \text{ Btu/lb}$$

Ex. 4.5. 1 ft³ of steam at 200 lbf/in² having a dryness fraction of 0·9 undergoes a non-flow constant volume process to a pressure of 83 lbf/in². Determine:

(a) The mass of steam involved
(b) The final dryness fraction
(c) The change in internal energy
(d) The change in enthalpy
(e) The heat energy transferred during the process, stating the direction of flow

Since values for steam at a pressure of 83 lbf/in² are not listed in the saturation tables, it is necessary to interpolate between the given values for 80 lbf/in² and 85 lbf/in².

Extract from saturation tables

p	t	v_g	u_f	u_g	h_f	h_{fg}	h_g
80	312	5·472	281·8	1,102·1	282	901·1	1,183·1
85	316·3	5·168	286·1	1,102·9	286·4	897·8	1,184·2

The values for 83 lbf/in² are now obtained by direct proportion, e.g.

$$t_{83} = t_{80} + \frac{83 - 80}{85 - 80} \times (t_{85} - t_{80})$$

$$= 312 + \frac{3}{5} \times (316 \cdot 3 - 312)$$

$$= 312 + 2 \cdot 58 = 314 \cdot 6°F$$

and

$$v_{g83} = v_{g80} + \frac{83 - 80}{85 - 80} \times (v_{g85} - v_{g80})$$

$$= 5 \cdot 472 + \frac{3}{5} \times (5 \cdot 168 - 5 \cdot 472)$$

$$= 5 \cdot 472 - 0 \cdot 182 = 5 \cdot 29 \text{ ft}^3/\text{lb, etc.}$$

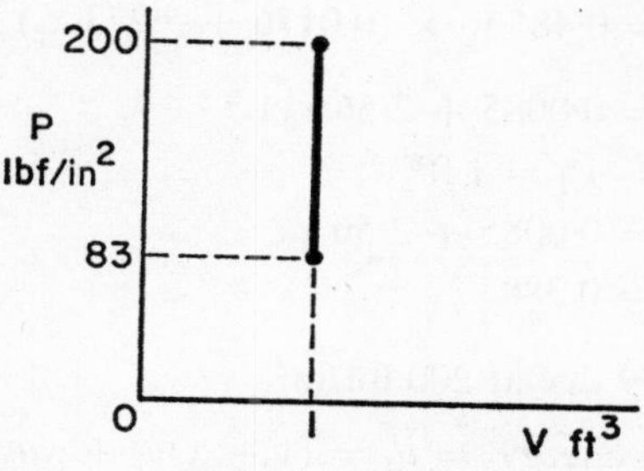

FIGURE E.4.1

Continuing the interpolation will give

p	t	v_g	u_f	u_g	h_f	h_{fg}	h_g
83	314·6	5·29	284·4	1,102·6	284·6	899·1	1,183·8

It should be noted that values of v_g and h_{fg} decrease for increasing values of pressure.

(a) Specific volume of steam 0·9 dry at 200 lbf/in² is given by

$$v_{x1} = (1 - x)v_f + xv_g$$

At 381·8°F the specific volume of water = 0·0184 ft³/lb

$$= (1 - 0·9) \times 0·0184 + 0·9 \times 2·288$$
$$= 0·00184 + 2·0592 = 2·061 \text{ ft}^3/\text{lb}$$

∴ Mass of 1 ft³ of steam 0·9 dry at 200 lbf/in² is

$$m = \frac{1 \text{ ft}^3}{2·061 \text{ ft}^3/\text{lb}} = 0·485 \text{ lb}$$

3

(b) Let the dryness fraction at 2 be x_2

Specific volume of steam with a dryness fraction x_2, at 83 lbf/in^2 is given by

$$v_{x2} = (1 - x_2)v_{f2} + x_2 v_{g2}$$

At 314·6°F the specific volume of water = 0·0176 ft^3/lb

$$= (1 - x_2) \times 0·0176 + x_2 \times 5·29$$
$$= 0·0176 + 5·273x_2 \text{ ft}^3/\text{lb}$$

$\therefore$ Volume of 0·485 lb of steam having a dryness fraction x_2 at 83 lbf/in^2 is given by

$$V_2 = mv_{x2}$$

$$= 0·485 \text{ lb} \times (0·0176 + 5·273 \, x_2) \frac{\text{ft}^3}{\text{lb}}$$

$$= 0·0085 + 2·56 \, x_2 \text{ ft}^3$$

But $\qquad V_2 = V_1 = 1 \text{ ft}^3$

$\therefore \qquad \qquad 1 = 0·0085 + 2·56 \, x_2$

$\therefore \qquad \qquad x_2 = 0·388$

(c) For steam 0·9 dry at 200 lbf/in^2

specific internal energy $= u_1 = (1 - x)u_f + xu_g$
$$= (1 - 0·9) \times 354·7 + 0·9 \times 1,113·7$$
$$= 35·47 + 1,002·33 = 1,037·8 \text{ Btu/lb}$$

For 0·485 lb, $U_1 = mu_1$

$$= 0·485 \text{ lb} \times 1,037·8 \frac{\text{Btu}}{\text{lb}}$$

$$= 503 \text{ Btu}$$

For steam 0·388 dry at 83 lbf/in^2

specific internal energy $= u_2 = (1 - x)u_f + xu_g$
$$= (1 - 0·388) \times 284·4 + 0·388 \times 1,102·6$$
$$= 174 + 428 = 602 \text{ Btu/lb}$$

$\therefore$ for 0·485 lb, $U_2 = mu_2$
$$= 0·485 \text{ lb} \times 602 \text{ Btu/lb}$$
$$= 292 \text{ Btu}$$

$\therefore$ change in internal energy $= U_2 - U_1$
$$= 292 - 503 = -211 \text{ Btu}$$

i.e. the internal energy decreases by 211 Btu during the process.

(d) For steam 0·9 dry at 200 lbf/in²

$$\begin{aligned}
\text{specific enthalpy} &= h_1 \\
&= (1 - x)h_f + xh_g \\
&= (1 - 0.9) \times 355.4 + (0.9 \times 1{,}198.4) \\
&= 35.54 + 1{,}078.56 = 1{,}114.1 \text{ Btu/lb}
\end{aligned}$$

$\therefore$ for 0·485 lb, $H_1 = mh_1$

$$= 0.485 \text{ lb} \times 1{,}114.1 \frac{\text{Btu}}{\text{lb}} = 540 \text{ Btu}$$

For steam 0·388 dry at 83 lbf/in²

$$\begin{aligned}
\text{specific enthalpy} &= h_2 \\
&= (1 - x)h_f + xh_g \\
&= (1 - 0.388) \times 284.6 + (0.388 \times 1{,}183.8) \\
&= 174 + 460 = 634 \text{ Btu/lb}
\end{aligned}$$

$\therefore$ for 0·485 lb, $H_2 = mh_2$

$$= 0.485 \text{ lb} \times 634 \frac{\text{Btu}}{\text{lb}}$$

$$= 307 \text{ Btu}$$

$\therefore$ change in enthalpy $= H_2 - H_1$
$$= 307 - 540 = -233 \text{ Btu}$$

i.e. the enthalpy decreases by 233 Btu during the process.

(e) Non-flow energy equation gives

$$Q - \frac{W}{J} = U_2 - U_1$$

For a constant volume process

$$\frac{W}{J} = 0$$

$$\therefore \quad Q = U_2 - U_1$$
$$= -211 \text{ Btu}$$

i.e. 211 Btu flow from the system to the surroundings during the process.

4.12 Measurement of Dryness Fraction

In fixing the state point and hence the properties of a wet mixture it was shown in §4.5 that the pressure and temperature alone were insufficient, as they are not independent properties in this region. If the dryness fraction is known however, the exact position of the state

APPLIED HEAT

58

point, and hence the values of the properties can be determined. The following methods of measuring the dryness fraction of a wet mixture have been devised.

4.13 Barrel Calorimeter

A barrel suitably insulated and containing water is arranged on the platform of a weighing machine. The temperature of the water is recorded when conditions have become steady. Steam at a known

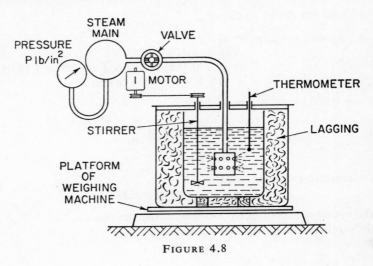

FIGURE 4.8

pressure P is allowed to pass into the water for a certain time, and the increase in temperature together with the increase in weight are recorded.

Let W_b = weight of the barrel
s = specific heat of the barrel material
t_1 = initial temperature of the water
t_2 = final temperature of the water
W_1 = initial weight of the water in the barrel
W_2 = final weight of the water in the barrel

Then weight of steam condensed = $W_2 - W_1$

Let P = steam pressure
h_f = specific enthalpy of saturated liquid } at pressure P
h_g = specific enthalpy of saturated vapour
x = dryness fraction of the wet steam

Then specific enthalpy of the wet steam $= (1 - x)h_f + xh_g$

Let $h_{f_1} =$ specific enthalpy of water at temperature t_1
and $h_{f_2} =$ specific enthalpy of water at temperature t_2

The heat energy transfer from the steam condensed (Q_1)

= weight of steam condensed × change in specific enthalpy
$= (W_2 - W_1) [(1 - x)h_f + xh_g - h_{f_2}]$

The heat energy transfer to the barrel and the water (Q_2)

= (weight of water + water equivalent of barrel) × change in specific enthalpy
$= (W_1 + W_b × s) (h_{f_2} - h_{f_1})$

Assuming there are no losses of heat energy, $Q_1 = Q_2$ and hence the dryness fraction x can be calculated.

Alternatively, initial enthalpy of system + enthalpy added = final enthalpy of system

$$(W_1 h_{f_1} + W_b s h_{f_1}) + (W_2 - W_1)\{(1 - x)h_f + xh_g\} = W_2 h_{f_2} + W_b s h_{f_2}$$

From this equation, x may be found as before.

4.14 Separating Calorimeter

A separating calorimeter is illustrated in Fig. 4.9. Wet steam is admitted at the top of the cylinder and is made to change direction before reaching the outlet at the upper end of the calorimeter. When the steam changes direction, the inertia of the relatively heavy droplets of water causes them to continue in the original direction and be collected in the graduated vessel. After the dry steam leaves the calorimeter, the rate of flow is obtained by either passing it through a calibrated nozzle or by condensing the steam and weighing.

During the same period of time

Let $W_1 =$ weight of water collected in the graduated vessel
$W_2 =$ weight of dry steam leaving the calorimeter

Then total weight of steam entering the calorimeter $= W_1 + W_2$

$\therefore$ dryness fraction of steam $= \dfrac{\text{weight of dry steam}}{\text{weight of wet mixture}}$

$$= \frac{W_2}{W_1 + W_2}$$

It should be noted that since mechanical separation is never perfect, some water droplets in suspension will leave the calorimeter with the dry steam. This will result in the apparent weight of dry steam being greater than the true weight, and hence the dryness fraction as given by this type of calorimeter is normally higher than the actual value.

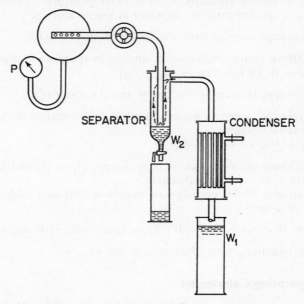

FIGURE 4.9

4.15 Throttling Calorimeter

The throttling calorimeter makes use of the fact that during a throttling process the enthalpy before throttling is equal to the enthalpy after throttling (see §2.11). Steam is allowed to pass through a small orifice where it must undergo a sufficiently large pressure drop to become superheated after the orifice. Readings of the temperature

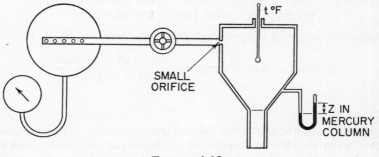

FIGURE 4.10

$t°F$ and pressure z inHg of the steam after throttling, the pressure of the steam in the main, and the barometric pressure z_b inHg are sufficient to enable the dryness fraction x to be determined.

Let h_f = specific enthalpy of saturated liquid $\}$ at the absolute
h_g = specific enthalpy of saturated vapour $\}$ pressure in the main

Then specific enthalpy before throttling = $(1 - x)h_f + xh_g$

$$\text{pressure after throttling} = (z + z_b) \text{ inHg}$$
$$= (z + z_b) \times \frac{14 \cdot 7}{30} \text{ lbf/in}^2$$

Temperature after throttling = $t°F$.

From these readings the specific enthalpy of the steam after throttling (h_s) can be obtained from the superheat tables.

Since

$$\text{enthalpy before throttling} = \text{enthalpy after throttling}$$
$$\therefore \quad (1 - x)h_f + xh_g = h_s$$

From this equation, the dryness fraction x can be determined. The use of the throttling calorimeter is limited by the fact that the steam must be superheated after throttling in order that the specific enthalpy may be determined by the pressure and temperature. Therefore, the limiting value of x for which the instrument can be used will be obtained when h_s is equal to the enthalpy of saturated vapour at the outlet conditions. In practice, this value of x is approximately 0·97. If the dryness fraction to be measured has a lower value than this, the following device may be used.

4.16 Combined Separating and Throttling Calorimeter

Steam from the mains is passed first through a separating calorimeter, where most of the water is removed. The exhaust steam from the separating calorimeter is then passed through a throttling calorimeter. During the same period of time let

W_1 = weight of water collected in separator
W_2 = weight of steam passing through the throttling calorimeter

Then

$W_1 + W_2$ = weight of wet steam entering separating calorimeter
Let h_f = specific enthalpy of saturated liquid $\}$ at mains pressure
h_g = specific enthalpy of saturated vapour $\}$
h_s = specific enthalpy of superheated steam after throttling

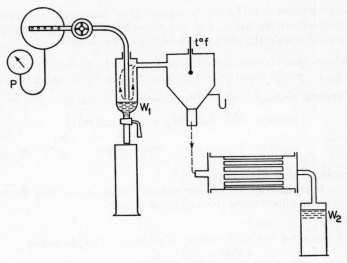

FIGURE 4.11

Enthalpy entering system = enthalpy leaving system

$$\therefore \quad (W_1 + W_2)\{h_f(1 - x) + xh_g\} = W_1h_f + W_2h_s$$

from which x can be determined.

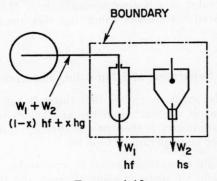

FIGURE 4.12

Ex. 4.6. In determining the dryness fraction of steam by means of a combined separating throttling calorimeter, the following observations were recorded.

Steam pressure in main 200 lbf/in^2
Steam pressure after throttling 14·7 lbf/in^2

Steam temperature after throttling 250°F
Rate of steam flow through throttle 4·8 lb/min
Rate of water collection in separator 0·5 lb/min

Determine:

(a) The dryness fraction of the steam in the main
(b) The dryness fraction of the steam between the separator and the throttle.

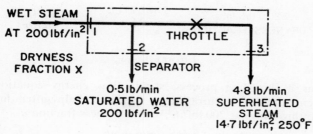

FIGURE E.4.2

For steady conditions

mass entering system per minute = mass leaving system per minute
∴ rate of steam flow into system = 0·5 lb/min + 4·8 lb/min
= 5·3 lb/min = m_1

(a) Let the dryness fraction of the steam in the main be x. Then specific enthalpy at $1 = h_1 = (1 - x)h_f + xh_g$.
From saturation tables, at 200 lbf/in²

$$h_f = 355·4 \text{ Btu/lb} = h_2$$
$$h_g = 1,198·4 \text{ Btu/lb}$$

From superheat tables, at 14·7 lbf/in² and 250°F

$$h_3 = 1,169·0 \text{ Btu/lb}$$

Applying the steady flow energy equation to the system gives

$$Q - \frac{W}{J} = [(H_2 + H_3) - H_1]$$

For a well-lagged system $Q = 0$. Also $W/J = 0$.

∴ $$H_2 + H_3 - H_1 = 0$$
i.e. $$m_2 h_2 + m_3 h_3 - m_1 h_1 = 0$$
∴ $$m_1 h_1 = m_2 h_2 + m_3 h_3$$

$$\therefore \quad 5 \cdot 3 \frac{lb}{min} \times [(1 - x)355 \cdot 4 + x1{,}198 \cdot 4] \frac{Btu}{lb}$$

$$= 0 \cdot 5 \frac{lb}{min} \times 355 \cdot 4 \frac{Btu}{lb} + 4 \cdot 8 \frac{lb}{min} \times 1{,}169 \cdot 0 \frac{Btu}{lb}$$

$$\therefore \quad 355 \cdot 4 - 355 \cdot 4x + 1{,}198 \cdot 4x = 33 \cdot 5 + 1{,}058$$

$$\therefore \quad x = \frac{736 \cdot 1}{843} = 0 \cdot 874$$

$$\underset{\substack{\text{DRYNESS FRACTION} \\ q}}{\overset{\text{200 lbf/in}^2}{\longrightarrow}} \quad \overset{\text{THROTTLE}}{\times} \quad \underset{250°F}{\text{14·7 lbf/in}^2}$$

FIGURE E.4.3

(b) For a throttling process, steady flow energy equation gives specific enthalpy before throttling = specific enthalpy after throttling. Before throttling, at 200 lbf/in² with a dryness fraction q

$$\text{specific enthalpy} = (1 - q)h_f + qh_g$$
$$= (1 - q)355 \cdot 4 + q1{,}198 \cdot 4 \text{ Btu/lb}$$

After throttling, at 14·7 lbf/in², 250°F, steam is superheated and specific enthalpy = 1,169·0 Btu/lb

$$\therefore \quad (1 - q)355 \cdot 4 + q1{,}198 \cdot 4 = 1{,}169$$

$$\therefore \quad q = \frac{1{,}169 - 355 \cdot 4}{1{,}198 \cdot 4 - 355 \cdot 4}$$

$$= \frac{813 \cdot 6}{843} = 0 \cdot 965$$

Ex. 4.7. 1 lb of saturated steam at 150 lbf/in² undergoes a non-flow constant volume process until the pressure becomes 50 lbf/in². Determine:

(a) The final condition of the steam
(b) The change in specific internal energy
(c) The change in specific enthalpy
(d) The heat energy transferred during the process, stating the direction of flow

Extracts from tables

p	t	v_g	u_f	u_g	h_f	h_{fg}	h_g
150	358·4	3·015	330·0	1,110·5	330·5	863·6	1,194·1
50	281	8·515	249·9	1,095·3	250·1	924	1,174·1

Since the steam is saturated at 1, the properties at 1 may be obtained direct from the saturation tables.

For a constant volume process

(*a*) At 2,

actual volume of 1 lb of steam = volume of steam at 1
= 3·015 ft^3
= volume of steam at 2

Since this is less than the specific volume of saturated steam at 2, then the steam at 2 must be wet.

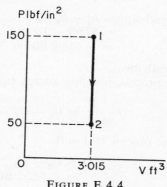

FIGURE E.4.4

If the dryness fraction at 2 is x

then
$$v_x = (1 - x)v_f + xv_g$$

i.e.
$$3·015 = (1 - x) \times 0·017 + x8·515$$
$$= 0·017 + x8·498$$

∴
$$x = \frac{2·998}{8·498} = 0·353$$

(*b*) At 1, where the steam is saturated at 150 lbf/in^2
specific internal energy = u_1 = 1,110·5 Btu/lb

At 2, where the steam is 0·353 dry at 50 lbf/in^2

specific internal energy = u_2
$$= (1 - x)u_f + xu_g$$
$$= (1 - 0·353) \times 249·9 + (0·353 \times 1,095·3)$$
$$= 161·8 + 386·5$$
$$= 548·3 \text{ Btu/lb}$$

∴ change in specific internal energy = $u_2 - u_1$
$$= 548·3 - 1,110·5$$
$$= -562·2 \text{ Btu/lb}$$

i.e. a loss of internal energy.

(c) At 1 for saturated steam at 150 lbf/in^2

specific enthalpy $= h_1 = h_g = 1{,}194 \cdot 1$ Btu/lb

At 2 for steam 0·353 dry at 50 lbf/in^2

$$\begin{aligned}
\text{specific enthalpy} = h_2 &= (1 - x)h_f + xh_g \\
&= (1 - 0\cdot353) \times 250\cdot1 + (0\cdot353 \times 1{,}174\cdot1) \\
&= 161\cdot9 + 414 \\
&= 575\cdot9 \text{ Btu/lb}
\end{aligned}$$

$$\begin{aligned}
\therefore \quad \text{change in specific enthalpy} &= h_2 - h_1 \\
&= 575\cdot9 - 1{,}194\cdot1 \\
&= -618\cdot2 \text{ Btu/lb}
\end{aligned}$$

i.e. a loss of specific enthalpy.

(d) For a non-flow process, non-flow energy equation gives

$$Q - \frac{W}{J} = U_2 - U_1$$

For a constant volume process $W/J = 0$

$$\therefore \qquad \begin{aligned} Q &= U_2 - U_1 \\ &= -562\cdot2 \text{ Btu/lb} \end{aligned}$$

i.e. 562·2 Btu/lb are transferred from the steam in the form of heat energy during the process.

Ex. 4.8. 0·5 lb of water at 100 lbf/in^2 and 60°F is contained in a cylinder 1 ft diameter by a frictionless piston. Heat energy is supplied until the temperature of the cylinder contents becomes 400°F, the pressure of the contents remaining at 100 lbf/in^2. Determine

(a) The heat energy supplied
(b) The distance moved by the piston
(c) The work energy transferred
(d) The change in internal energy

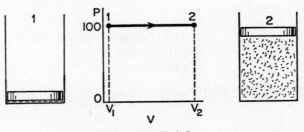

FIGURE E.4.5

(a) During a constant pressure process, the heat energy supplied to a fluid is equal to the change in enthalpy.

State 1. 0.5 lb of water at 100 lbf/in^2 and 60°F.

From the saturation tables, the temperature corresponding to a pressure of 100 lbf/in^2 is 327·8°F. Since the water is at 60°F, i.e. at a temperature below that corresponding to the pressure, the properties of the water are taken as those of water at 60°F and the pressure corresponding to 60°F, i.e. 0·2563 lbf/in^2 (from tables), i.e.

$$h_1 = 28·1 \text{ Btu/lb}$$

and for 0.5 lb of water

$$
\begin{aligned}
H_1 &= mh_1 \\
&= 0·5 \text{ lb} \times 28·1 \text{ Btu/lb} \\
&= 14·05 \text{ Btu}
\end{aligned}
$$

State 2. 0.5 lb of fluid at 100 lbf/in^2, 400°F.

Since the temperature of the cylinder contents is above that corresponding to the pressure, i.e. above 327·8°F the cylinder must contain superheated steam at 2, and the properties may be obtained from the superheat tables, i.e. at 100 lbf/in^2 and 400°F, $h_2 = 1,227·6$ Btu/lb

and for 0.5 lb,
$$
\begin{aligned}
H_2 &= mh_2 \\
&= 0·5 \text{ lb} \times 1,227·6 \text{ Btu/lb} \\
&= 613·8 \text{ Btu}
\end{aligned}
$$

$\therefore$ heat energy supplied $= H_2 - H_1$
$$
\begin{aligned}
&= 613·8 \text{ Btu} - 14·05 \text{ Btu} \\
&= 599·75 \text{ Btu}
\end{aligned}
$$

(b) Initial specific volume of water at 60°F = 0·016 ft^3/lb

$\therefore$ volume of 0.5 lb $= 0·5$ lb $\times$ 0·016 ft^3/lb
$$= 0·008 \text{ ft}^3$$

specific volume of superheated steam at 100 lbf/in^2 and 400°F
$$= 4·937 \text{ ft}^3\text{/lb}$$

$\therefore$ volume of 0.5 lb $= 0·5$ lb $\times$ 4·937 ft^3/lb
$$= 2·4685 \text{ ft}^3$$

$\therefore$ change in volume $= 2·4685$ ft^3 $-$ 0·008 ft^3
$$= 2·4605 \text{ ft}^3$$

$\therefore$ distance moved by piston $= \dfrac{\text{volume change}}{\text{cylinder area}}$
$$= \frac{2·4605 \text{ ft}^3}{\frac{1}{4}\pi \times 1^2 \text{ ft}^2} = 3·135 \text{ ft}$$

(c) For a constant pressure process,

$$\text{work energy transferred} = \frac{W}{J} = \frac{P(V_2 - V_1)}{J}$$

$$= \frac{(100 \times 144)(\text{lbf/ft}^2) \times 2 \cdot 4605 \text{ ft}^3}{778 \text{ ft lbf/Btu}}$$

$$= 45 \cdot 5 \text{ Btu}$$

(d) Non-flow energy equation gives

$$Q - \frac{W}{J} = U_2 - U_1$$

∴ $$599 \cdot 75 - 45 \cdot 5 = U_2 - U_1$$
∴ $$U_2 - U_1 = 554 \cdot 25 \text{ Btu}$$

i.e. a gain in internal energy.

Ex. 4.9. 5 ft³ of steam are contained in a cylinder at 150 lbf/in² and 450°F. If 800 Btu of heat energy flow from the steam to the surroundings whilst the pressure of the steam remains constant, determine

(a) The final condition of the steam
(b) The work energy transferred
(c) The change in internal energy during the process

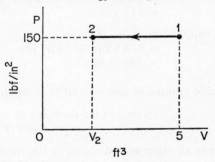

FIGURE E.4.6

Since the mass of steam in the cylinder is not given, it must be calculated.
At 1, at 150 lbf/in² and 450°F, steam is superheated.
From superheat tables

$$\text{specific volume at 150 lbf/in}^2 \text{ and 450°F} = 3 \cdot 457 \text{ ft}^3/\text{lb}$$

∴ $$\text{mass of 5 ft}^3 \text{ of this steam} = \frac{5 \text{ ft}^3}{3 \cdot 457 \text{ ft}^3/\text{lb}}$$

$$= 1 \cdot 448 \text{ lb}$$

(a) For a constant pressure process, heat energy transferred is equal to the change in enthalpy.

From superheat tables

$$h_1 = 1{,}247 \cdot 5 \text{ Btu/lb}$$

$$\therefore \qquad H_1 = mh_1$$

$$= 1 \cdot 448 \text{ lb} \times 1{,}247 \cdot 5 \,\frac{\text{Btu}}{\text{lb}}$$

$$= 1{,}805 \text{ Btu}$$

Since $\qquad\qquad Q = H_2 - H_1$

then $\qquad -800 \text{ Btu} = H_2 - 1{,}805 \text{ Btu}$

$$\therefore \qquad\qquad H_2 = 1{,}005 \text{ Btu}$$

and $\qquad\qquad h_2 = \dfrac{H_2}{m} = \dfrac{1{,}005 \text{ Btu}}{1 \cdot 448 \text{ lb}}$

$$= 695 \text{ Btu/lb}$$

From saturation tables at 150 lbf/in², $h_f = 330 \cdot 5$ Btu/lb and $h_g = 1{,}194 \cdot 1$ Btu/lb. Since h_2 lies between h_f and h_g, the steam at 2 must be wet.

Let the dryness fraction at $2 = x$.

Then $\qquad\qquad h_2 = (1 - x)h_f + xh_g$

i.e. $\qquad 695 = (1 - x)330 \cdot 5 + x1{,}194 \cdot 1$

$$\therefore \qquad\qquad x = \frac{695 - 330 \cdot 5}{1{,}194 \cdot 1 - 330 \cdot 5}$$

$$= \frac{364 \cdot 5}{863 \cdot 6} = 0 \cdot 422$$

i.e. the steam at 2 has a dryness fraction of 0·422.

(b) For a constant pressure process, the work energy transferred is given by

$$\frac{W}{J} = \frac{P(V_2 - V_1)}{J}$$

At 1 $\qquad\qquad V_1 = 5 \text{ ft}^3$

At 2, specific volume of steam 0·422 dry at 150 lbf/in² is

$$v_x = (1 - x)v_f + xv_g$$
$$= (1 - 0 \cdot 422) \times 0 \cdot 016 + (0 \cdot 422 \times 3 \cdot 015)$$
$$= 0 \cdot 00925 + 1 \cdot 272 = 1 \cdot 281 \text{ ft}^3/\text{lb}$$

$$\therefore \qquad \text{final volume of steam} = V_2 = mv_2$$
$$= 1 \cdot 448 \text{ lb} \times 1 \cdot 281 \text{ ft}^3/\text{lb}$$
$$= 1 \cdot 854 \text{ ft}^3$$

$$\therefore \qquad \frac{W}{J} = \frac{(150 \times 144)(\text{lbf/ft}^2) \times (1 \cdot 854 - 5) \text{ ft}^3}{778 \text{ ft lbf/Btu}}$$

$$= -87 \cdot 3 \text{ Btu}$$

i.e. 87·3 Btu of work energy are transferred to the steam.

(c) Non-flow energy equation gives

$$Q - \frac{W}{J} = U_2 - U_1$$

i.e. $\qquad -800 - (-87 \cdot 3) = U_2 - U_1$

$\therefore \qquad U_2 - U_1 = -712 \cdot 7 \text{ Btu}$

i.e. the internal energy of the steam decreases by 712·7 Btu.

Ex. 4.10. 2 ft³ of dry saturated steam at 120 lbf/in² are contained in a cylinder by a frictionless piston. If the steam undergoes hyperbolic expansion to a pressure of 60 lbf/in², determine

(a) The work energy transferred
(b) The change in internal energy
(c) The heat energy transferred

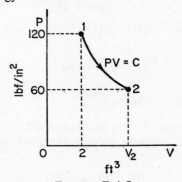

FIGURE E.4.7

Since the expansion is hyperbolic

$$P_1V_1 = P_2V_2$$

i.e. $\qquad (120 \times 144)\dfrac{\text{lbf}}{\text{ft}^2} \times 2 \text{ ft}^3 = (60 \times 144)\dfrac{\text{lbf}}{\text{ft}^2} \times V_2$

$\therefore \qquad V_2 = 4 \text{ ft}^3$

(a) For a hyperbolic process

$$\frac{W}{J} = \frac{P_1 V_1}{J} \ln \frac{V_2}{V_1} \text{ (see Appendix 2)}$$

$$= \frac{(120 \times 144) \frac{\text{lbf}}{\text{ft}^2} \times 2 \text{ ft}^3}{778 \frac{\text{ft lbf}}{\text{Btu}}} \log_e \left(\frac{4 \text{ ft}^3}{2 \text{ ft}^3}\right)$$

$$= 30 \cdot 8 \text{ Btu}$$

(b) To determine the internal energies, the mass of steam in the cylinder must first be determined, together with the conditions of the steam.

At 1, steam is saturated at 120 lbf/in².
From saturation tables, at 120 lbf/in²,

specific volume of saturated steam = 3·728 ft³/lb

∴ mass of steam in cylinder $= \dfrac{\text{volume of steam}}{\text{volume per lb}}$

$$= \frac{2 \text{ ft}^3}{3 \cdot 728 \text{ (ft}^3/\text{lb)}}$$

$$= 0 \cdot 536 \text{ lb}$$

From saturation tables, at 120 lbf/in²
specific internal energy of saturated steam = u_1 = 1,107·6 Btu/lb

∴ $U_1 = m u_1$

$$= 0 \cdot 536 \text{ lb} \times 1,107\ 6 \frac{\text{Btu}}{\text{lb}}$$

$$= 593 \text{ Btu}$$

To determine the condition of the steam at 2

From saturation tables, at 60 lbf/in²

specific volume of saturated steam = $7 \cdot 175 \dfrac{\text{ft}^3}{\text{lb}}$

∴ volume of 0·536 lb of saturated steam $= 7 \cdot 175 \dfrac{\text{ft}^3}{\text{lb}} \times 0 \cdot 536 \text{ lb}$

$$= 3 \cdot 85 \text{ ft}^3$$

Since the volume of the steam at 2 is greater than 3·85 ft³, then the steam at 2 must be superheated.
The internal energy of superheated steam may be determined by

first calculating the enthalpy of the steam, and then applying the expression

$$h = u + \frac{pv}{J}$$

The enthalpy of the superheated steam may be obtained either by interpolation from the superheat tables (since the pressure and specific volume are known) or by applying the formula (§ 4.10)

$$pv = 1 \cdot 247(h - 835)$$

which is obtainable in the steam tables. It should be noted that in this formula, p is the pressure in lbf/in², v is the specific volume in ft³/lb, and h is the specific enthalpy in Btu/lb.

At 2
$$p = 60 \text{ lbf/in}^2$$
$$v = \frac{4 \text{ ft}^3}{0 \cdot 536 \text{ lb}} = 7 \cdot 45 \text{ ft}^3/\text{lb}$$

$$\therefore \qquad 60 \times 7 \cdot 45 = 1 \cdot 247(h_2 - 835)$$

$$\therefore \qquad h_2 = 359 + 835 = 1{,}194 \text{ Btu/lb}$$

$$\therefore \quad u_2 = h_2 - \frac{p_2 v_2}{J}$$
$$= 1{,}194 \text{ Btu/lb} - \frac{(60 \times 144) \text{ lbf/ft}^2 \times 7 \cdot 45 \text{ ft}^3/\text{lb}}{778 \text{ ft lbf/Btu}}$$
$$= 1{,}194 - 82 \cdot 7$$
$$= 1{,}111 \cdot 3 \text{ Btu/lb}$$

$$\therefore \; U_2 = mu_2$$
$$= 0 \cdot 536 \text{ lb} \times 1{,}111 \cdot 3 \; \frac{\text{Btu}}{\text{lb}}$$
$$= 597 \text{ Btu}$$

$$\therefore \; \text{change in internal energy} = U_2 - U_1$$
$$= 597 - 593$$
$$= 4 \text{ Btu}$$

(c) The heat energy transferred during the process is obtained from the non-flow energy equation

$$Q - \frac{W}{J} = U_2 - U_1$$

$$\therefore \qquad Q - 30 \cdot 8 \text{ Btu} = 4 \text{ Btu}$$

$$\therefore \qquad Q = 34 \cdot 8 \text{ Btu}$$

i.e. 34·8 Btu are supplied during the process.

Note: When the process is hyperbolic, i.e. it follows the law $PV =$ constant, the following method is available for determining the change in internal energy.

Since
$$H = U + \frac{PV}{J}$$

then
$$H_1 = U_1 + \frac{P_1 V_1}{J}$$

and
$$H_2 = U_2 + \frac{P_2 V_2}{J}$$

from which

$$H_2 - H_1 = (U_2 - U_1) + \left(\frac{P_2 V_2}{J} - \frac{P_1 V_1}{J}\right)$$
$$= U_2 - U_1$$

since in this case, $P_1 V_1 = P_2 V_2$.

It must be emphasised that this alternative method is only available for a hyperbolic process.

Ex. 4.11. 1 lb of steam at 250 lbf/in² and 450°F undergoes a non-flow process to a pressure of 100 lbf/in². If the final volume of the steam is 4·5 ft³, determine

(a) The equation of the process
(b) The work energy transferred
(c) The final temperature of the steam

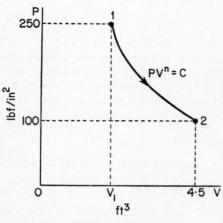

FIGURE E.4.8

(d) The change in internal energy
(e) The heat energy transferred
(f) The change in enthalpy

From the tables for superheated steam at 250 lbf/in² and 450°F,

$$\text{specific volume} = 2{\cdot}002 \text{ ft}^3/\text{lb} = V_1$$

(a) Assuming the process to follow the law $PV^n = \text{constant}$

$$P_1V_1{}^n = P_2V_2{}^n$$

i.e. $\quad (250 \times 144)\dfrac{\text{lbf}}{\text{ft}^2} \times (2{\cdot}002 \text{ ft}^3)^n = (100 \times 144)\dfrac{\text{lbf}}{\text{ft}^2} \times (4{\cdot}5 \text{ ft}^3)^n$

$$\therefore \quad \frac{250 \times 144}{100 \times 144} = \left(\frac{4{\cdot}5}{2{\cdot}002}\right)^n = 2{\cdot}25^n$$

$$\therefore \quad\quad\quad n = 1{\cdot}13$$

$\therefore$ equation of process is $PV^{1{\cdot}13} = \text{constant}$.

(b) The work energy transferred during the process is given by

$$\frac{W}{J} = \frac{P_1V_1 - P_2V_2}{J(n-1)}$$

$$= \frac{(250 \times 144) \text{ lbf/ft}^2 \times 2{\cdot}002 \text{ ft}^3 - (100 \times 144) \text{ lbf/ft}^2 \times 4{\cdot}5 \text{ ft}^3}{778 \text{ ft lbf/Btu} \times (1{\cdot}13 - 1)}$$

$$= \frac{72{,}050 - 64{,}800}{778 \times 0{\cdot}13} \text{ Btu}$$

$$= 71{\cdot}7 \text{ Btu}$$

(c) Final volume of 1 lb of steam $= 4{\cdot}5 \text{ ft}^3$
From the tables for saturated steam, the specific volume of saturated steam at 100 lbf/in² $= 4{\cdot}432$ ft³/lb. Since this is less than the actual final specific volume of the steam, then the steam at 2 must be superheated. The temperature of the steam at 2 may be obtained by interpolation from the steam tables as follows:

From saturation tables

specific volume of saturated steam at 100 lbf/in²(327·8°F) $= 4{\cdot}432$ ft³/lb

From superheat tables

specific volume of superheated steam at 100 lbf/in² (350°F) $= 4{\cdot}592$ ft³/lb

$\therefore$ Temperature of steam at 100 lbf/in² having a specific volume of 4·5 ft³/lb is given by

$$327 \cdot 8 + (350 - 327 \cdot 8) \times \frac{4 \cdot 5 - 4 \cdot 432}{4 \cdot 592 - 4 \cdot 432}$$

$$= 327 \cdot 8 + 22 \cdot 2 \times \frac{0 \cdot 068}{0 \cdot 160}$$

$$= 327 \cdot 8 + 9 \cdot 44$$

$$= 337 \cdot 24°F$$

(d) Since the steam is superheated throughout, the initial and final internal energies cannot be obtained directly from the normal tables. They can, however, be obtained from the expression for enthalpy

$$h = u + \frac{pv}{J}$$

At 1, $p = 250$ lbf/in², $t = 450°F$, $v = 2 \cdot 002$ ft³/lb

$\therefore$ from superheat tables

$$h_1 = 1{,}233 \cdot 5 \text{ Btu/lb}$$

$\therefore$ $u_1 = h_1 - \dfrac{p_1 v_1}{J}$

$$= 1{,}233 \cdot 5 \frac{\text{Btu}}{\text{lb}} - \frac{(250 \times 144) \text{ lbf/ft}^2 \times 2 \cdot 002 \text{ ft}^3/\text{lb}}{778 \text{ ft lbf/Btu}}$$

$$= 1{,}233 \cdot 5 \text{ Btu/lb} - 92 \cdot 6 \text{ Btu/lb}$$

$$= 1{,}140 \cdot 9 \text{ Btu/lb}$$

At 2, $p = 100$ lbf/in², $t = 337 \cdot 24°F$, $v = 4 \cdot 5$ ft³/lb

h_2 may be obtained by interpolation from the superheat tables, or alternatively by use of the expression

$$pv = 1 \cdot 247 \, (h - 835)$$

quoted in the steam tables. It should be noted that in this expression, p is the pressure in lbf/in², v the specific volume in ft³/lb and h the specific enthalpy in Btu/lb

$\therefore$ $p_2 v_2 = 1 \cdot 247 \, (h_2 - 835)$

i.e. $100 \times 4 \cdot 5 = 1 \cdot 247 \, (h_2 - 835)$

$\therefore$ $h_2 = 361 + 835$

$$= 1{,}196 \text{ Btu/lb}$$

$$\therefore \quad u_2 = h_2 - \frac{p_2 v_2}{J}$$

$$= 1,196 \frac{\text{Btu}}{\text{lb}} - \frac{(100 \times 144) \text{ lbf/ft}^2 \times 4.5 \text{ ft}^3/\text{lb}}{778 \text{ ft lbf/Btu}}$$

$$= 1,196 \text{ Btu/lb} - 83.3 \text{ Btu/lb}$$

$$= 1,112.7 \text{ Btu/lb}$$

$$\therefore \quad \text{change in internal energy} = u_2 - u_1$$
$$= 1,112.7 \text{ Btu/lb} - 1,140.9 \text{ Btu/lb}$$
$$= -28.2 \text{ Btu/lb}$$

(e) The heat energy transferred during the process is obtained by applying the non-flow energy equation

$$Q - \frac{W}{J} = u_2 - u_1$$

$$\therefore \quad Q - 71.7 \text{ Btu} = -28.2 \text{ Btu/lb}$$

$$\therefore \quad Q = 71.7 - 28.2 = +43.5 \text{ Btu/lb}$$

i.e. 43.5 Btu are supplied during the process.

(f) From (d)

$$h_1 = 1,233.5 \text{ Btu/lb}$$
$$h_2 = 1,196 \text{ Btu/lb}$$
$$\therefore \quad \text{change in enthalpy} = h_2 - h_1$$
$$= 1,196 - 1,233.5$$
$$= -37.5 \text{ Btu/lb}$$

EXERCISES ON CHAPTER 4

N.B. All pressures are absolute.

1. Steam at 50 lbf/in² has a dryness fraction of 0·8. Calculate its specific volume, specific enthalpy, and specific internal energy.

(6·185 ft³/lb; 989·9 Btu/lb; 926·2 Btu/lb)

2. A cylinder contains 0·75 lb of steam at 150 lbf/in². If the volume of the cylinder is 1·8 ft³, calculate the dryness fraction of the steam, the internal energy, and the enthalpy.

(0·795; 713 Btu; 762 Btu)

3. Calculate the volume, enthalpy, and internal energy of 2 lb of steam at 175 lbf/in² and 700°F.

(7·912 ft³; 2,750 Btu; 2,494 Btu)

4. A rigid vessel having a volume of 6 ft³ contains steam 0·9 dry at 120 lbf/in². If cooling takes place until the pressure becomes 40 lbf/in², calculate the mass of steam in the cylinder, its final dryness fraction, and the heat energy transferred.

(1·79 lb; 0·319; −929 Btu)

5. Calculate the heat energy required to convert 100 lb of water at 20 lbf/in² and 60°F into steam at 300°F during a constant pressure process.

(116,350 Btu)

6. A cylinder contains 10 ft³ of steam at 60 lbf/in² and 400°F. If the steam is cooled at constant pressure until the volume becomes 2 ft³, calculate the final dryness fraction, the work energy transfer, and the heat energy transfer.

(0·231; −88·8 Btu; −910 Btu)

7. During a non-flow hyperbolic process, 1 lb of steam initially 0·9 dry at 180 lbf/in² is expanded to a pressure of 40 lbf/in². Calculate the final dryness fraction, the work energy transfer, and the heat energy transfer.

(0·977; +114 Btu; +151·5 Btu)

8. A cylinder contains 1 ft³ of steam 0·8 dry at 200 lbf/in². The steam expands during a non-flow hyperbolic process until the volume becomes 8 ft³. Calculate the final pressure, the final dryness fraction, the work energy transfer, and the heat energy transfer.

(25 lbf/in²; 0·902; +77 Btu; +97·45 Btu)

9. 1 lb of steam 0·8 dry at 140 lbf/in² expands during a non-flow polytropic process according to the law $PV^{1·1}$ = constant until the pressure becomes 40 lbf/in². Calculate the final dryness fraction, the work energy transfer, and the heat energy transfer.

(0·767; +54,500 ft lbf; +13·8 Btu)

10. The cylinder of a steam engine contains 0·75 ft³ of dry saturated steam at 120 lb/in². If the steam undergoes polytropic expansion to a pressure of 50 lbf/in² and a volume of 1·6 ft³, calculate

(a) The mass of steam in the cylinder
(b) The final dryness fraction
(c) The index of expansion
(d) The change of internal energy
(e) The work energy transfer
(f) The heat energy transfer

(0·201 lb; 0·934; 1·155; −13·7 Btu; +9,300 ft lbf; −1·75 Btu)

5: Perfect Gases

One important type of fluid which has many applications in thermodynamics is the type in which the working temperature of the fluid remains well above the critical temperature of the fluid. In this case the fluid cannot be liquefied by an isothermal compression, i.e. if it is required to condense the fluid, then cooling of the fluid must first be carried out (see previous chapter, §4.3). In the simple treatment of such fluids, their behaviour is likened to that of a perfect gas. Although, strictly speaking, a perfect gas is an ideal which can never be realized in practice, the behaviour of many 'permanent' gases, e.g. hydrogen, oxygen, air, etc., is very similar to the behaviour of a perfect gas to a first approximation. Two of the laws describing the behaviour of a perfect gas, namely Boyle's Law and Charles' Law will already be known to the student.

5.1 Boyle's Law

This Law may be stated as follows: Provided the temperature T of a perfect gas remains constant, then the volume V of a given mass of gas

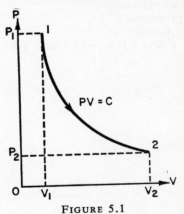

FIGURE 5.1

is inversely proportional to the pressure P of the gas, i.e. $P \propto 1/V$, or $P \times V = $ constant if temperature remains constant. If a gas changes from a state 1 to a state 2 during an isothermal process, then $P_1 \times V_1 = P_2 \times V_2 = $ a constant.

If the process is represented on a graph having axes of pressure P and volume V, the result will be as shown in Fig. 5.1. The curve is known as a rectangular hyperbola, having the mathematical equation, $xy =$ constant.

5.2 Charles' Law

This may be stated as follows: Provided the pressure P of a given mass of gas remains constant, then the volume V of the gas will be directly proportional to the absolute temperature T of the gas, i.e.

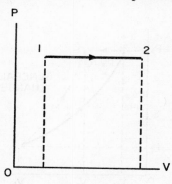

FIGURE 5.2

$V \propto T$, or $V =$ constant $\times T$. Therefore $V/T =$ constant, for constant pressure P.

If a gas changes from state 1 to state 2 during a constant pressure process, then $V_1/T_1 = V_2/T_2 =$ constant.

If the process is represented on a P–V diagram as before, the result will be as shown in Fig. 5.2.

5.3 General Change

Consider now a quantity of gas which is at state 1, i.e. it has a pressure P_1, volume V_1, absolute temperature T_1. Let the gas undergo a process after which it is at any state 2, i.e. the pressure is P_2, volume V_2, absolute temperature T_2.

On a P–V diagram this will appear as shown by the dotted line in Fig. 5.3. Bearing in mind that the change in any property depends only on the initial and final states, and not on the way these changes are brought about, the gas could be made to change from 1 to 2 by a combination of:

(a) An isothermal compression from state 1 to state x at temperature T_1, where $P_x = P_2$

This change in state would therefore obey Boyle's Law, i.e.

$$P_1V_1 = P_xV_x = P_2V_x$$

for constant temperature T_1 or

$$V_x = \frac{P_1V_1}{P_2} \tag{5.1}$$

(b) By keeping the pressure constant, and increasing the volume of the gas until the temperature becomes T_2. Since this is a constant

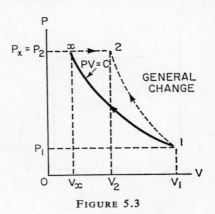

FIGURE 5.3

pressure process, it will obey Charles' Law, i.e. $V_x/T_x = V_2/T_2$ for constant pressure P_2.

Since $T_x = T_1$ this may be written

$$\frac{V_x}{T_1} = \frac{V_2}{T_2}$$

or

$$V_x = \frac{V_2}{T_2}T_1 \tag{5.2}$$

but also $V_x = \dfrac{P_1V_1}{P_2}$

$$\therefore \qquad \frac{P_1V_1}{P_2} = \frac{V_2T_1}{T_2}$$

or

$$\frac{P_1V_1}{T_1} = \frac{P_2V_2}{T_2} \tag{5.3}$$

Since 2 represented any state, then it will be true to say that P_1V_1/T_1 will be equal to PV/T at any other state of the gas, i.e. $PV/T = $ constant k.

If the mass of gas involved in the process is m lb, then dividing each side of the equation by m gives

$$\frac{PV}{mT} = \frac{k}{m} = \text{constant } (R)$$

R is called the Characteristic Gas Constant.

The equation $PV/mT = R$ is called the Characteristic Gas Equation and is normally written

$$PV = mRT \qquad (5.4)$$

The units of R will be

$$\frac{\text{pressure} \times \text{volume}}{\text{mass} \times \text{temperature}} = \frac{\text{lbf}}{\text{ft}^2} \times \frac{\text{ft}^3}{\text{lb degR}} = \frac{\text{ft lbf}}{\text{lb degR}}$$

A quantity of gas whose mass is equal to the molecular weight of the gas in pounds is called a 'pound mole' or sometimes just 'mol'. If the equation is applied to a pound mole of a gas of molecular weight M, the volume of the gas will now be the molar volume V_m and the mass of gas will be M lb. The characteristic equation will then become

$$PV_M = MRT$$

or

$$\frac{PV_M}{T} = MR \qquad (5.5)$$

It can be shown that for all gases at the same pressures and temperatures, the molar volume of each gas is the same (see §9.4), i.e. V_M is equal for all gases at one pressure and temperature. Equation (5.5), therefore, becomes

$$\frac{PV_M}{T} = \text{constant} = MR \text{ for all gases}$$

The constant MR is called the Universal Gas Constant, since it will have the same value for all gases. It is given the symbol R_0. Experiment shows that at 14·7 lbf/in² and 32°F the molar volume of any gas is 359 ft³

$$\therefore R_0 = MR = \frac{PV_M}{T} = \frac{(14\cdot7 \times 144) \times 359}{(460 + 32)} \frac{\text{lbf}}{\text{ft}^2} \frac{\text{ft}^3}{\text{mol}} \text{degR}$$

$$= 1{,}545 \text{ ft lbf/mol degR}$$

$$= 1\cdot986 \text{ Btu/mol degR}$$

If the molecular weight of a gas is known, then the Characteristic Gas Constant may be found by dividing R_0 by the molecular weight.

E.g. for O_2, the molecular weight is 32. Therefore characteristic

$$\text{constant} = \frac{1\cdot986 \times 778}{32} = 48\cdot2 \ \frac{\text{ft lbf}}{\text{lb degR}}$$

5.4 Specific Heats

The specific heat of any substance is defined as the amount of heat energy required to raise the temperature of unit mass of the substance through one degree, under specified conditions. In thermodynamics, two specified conditions are used, those of constant volume and constant pressure. If the working fluid is in the liquid state, e.g. water, the specific heat at constant volume will be much the same as the specific heat at constant pressure (since liquids normally have a low coefficient of expansion) and it is usual to use one common value for both. With gases however, the two specific heats do not have the same value and it is essential to distinguish between them.

If 1 lb of a gas is supplied with an amount of heat energy sufficient to raise the temperature of the gas by 1 degree whilst the volume of the gas remains constant, then the amount of heat energy supplied is known as the specific heat at constant volume, and is denoted by c_v. If 1 lb of a gas is supplied with a quantity of heat energy sufficient to raise the temperature of the gas by 1 degree whilst the pressure is held constant, then the amount of heat energy supplied is known as the specific heat at constant pressure and is denoted by c_p. For all elementary calculations it may be assumed that both c_p and c_v remain constant.

The applications of the energy equation to various non-flow processes which may be undergone by a perfect gas will now be considered.

5.5 Constant Volume Process

Consider m lb of a gas contained in a cylinder at temperature T_1, to which heat energy is supplied reversibly until the temperature becomes T_2; if the volume of the gas is held constant, then no work energy transfers can take place since the piston will be unable to move. If c_v = specific heat at constant volume, then the heat energy supplied = mass × specific heat at constant volume × temperature change.
i.e.

$$Q = m \times c_v \times (T_2 - T_1) \qquad (5.5)$$

Also, the non-flow energy equation gives

$$Q - W/J = U_2 - U_1$$
$$\therefore \qquad mc_v(T_2 - T_1) - 0 = U_2 - U_1$$
$$\text{or} \qquad mc_v(T_2 - T_1) = U_2 - U_1 \qquad (5.6)$$

i.e. all the heat energy transfers taking place in a constant volume process affect the internal energy only, and for a perfect gas, the change in internal energy during a constant volume process is equal to a constant × temperature change. Joule discovered experimentally (and this fact may also be verified mathematically) that the change of

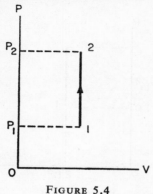

FIGURE 5.4

internal energy of a gas during any process depends only on the change in temperature of the gas. This is known as Joule's Law. Since the above expression for the change in internal energy is of this form (i.e. it depends only on the change of temperature) it will be true to say that for a gas undergoing any process, the change in internal energy will be given by $mc_v(T_2 - T_1)$.

Ex. 5.1. Two cubic feet of oxygen are contained in a sealed cylinder at a pressure of 20 lbf/in² and a temperature of 70°F. Determine the quantity of heat energy which must be supplied to the oxygen to increase its temperature to 200°F, and the pressure of the oxygen at the final conditions.

Given:
Universal gas constant is 1·986 Btu/mol degR
Molecular weight of oxygen = 32
c_v for oxygen = 0·236 Btu/lb degR
Since the cylinder is sealed, the process will take place under constant volume conditions.

From equation (5.5), for a constant volume process

heat energy supplied = $m \times c_v \times (T_2 - T_1)$

In this equation, c_v, T_1, and T_2 are given.
To determine m, equation (5.4) gives

$$m = \frac{PV}{RT}$$

where
$$R = \frac{R_0}{\text{mol. wt}} = \frac{1 \cdot 986}{32} \frac{\text{Btu}}{\text{mol degR}} \frac{\text{mol}}{\text{lb}}$$
$$= 0 \cdot 062 \text{ Btu/lb degR}$$
$$= 48 \cdot 2 \text{ ft lbf/lb degR}$$

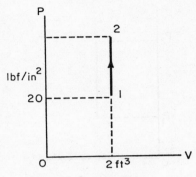

FIGURE E.5.1

$\therefore$ at
$$1 \, m = \frac{P_1 V_1}{R T_1}$$

$$= \frac{(20 \times 144) \text{ lbf/ft}^2 \times 2 \text{ ft}^3}{48 \cdot 2 \text{ ft lbf/lb degR} \times (460 + 70) \text{ degR}}$$

$$= 0 \cdot 2255 \text{ lb}$$

$\therefore$ heat energy supplied $= m \times c_{\text{v}} \times (T_2 - T_1)$

$$= 0 \cdot 2255 \text{ lb} \times 0 \cdot 236 \frac{\text{Btu}}{\text{lb degR}} \times [(460 + 200)$$
$$-(460 + 70)] \text{ degR}$$

$$= 6 \cdot 92 \text{ Btu}$$

To find the final pressure P_2, equation (5.3) gives

$$\frac{P_1 V_1}{T_1} = \frac{P_2 V_2}{T_2}$$

$\therefore$
$$P_2 = P_1 \frac{T_2}{T_1} \text{ since } V_1 = V_2 = 2 \text{ ft}^3$$

$$= 20 \text{ lbf/in}^2 \times \frac{(460 + 200)}{(460 + 70)} \frac{°\text{R}}{°\text{R}}$$

$$= 24 \cdot 9 \text{ lbf/in}^2$$

5.6 Constant Pressure Process

Consider m lb of a gas at temperature T_1 having a volume V_1 contained in a cylinder by a piston under a constant pressure P. Let heat energy be supplied to the gas until the temperature of the gas rises to

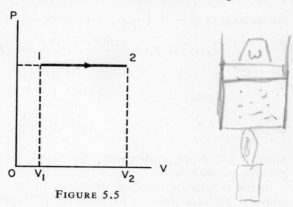

FIGURE 5.5

T_2. As heat energy is supplied, the gas will expand and push the piston outwards until the gas occupies a volume V_2, i.e. some energy will be transferred from the gas to the surroundings in the form of work energy by the movement of the piston. The amount of work energy transferred will be given by

$$\frac{W}{J} = \frac{P(V_2 - V_1)}{J} \text{ heat units}$$

Heat energy supplied = mass × specific heat at constant pressure × temperature change.

$$Q = m \times c_\mathrm{p} \times (T_2 - T_1) \tag{5.8}$$

The non-flow energy equation gives

$$Q - \frac{W}{J} = U_2 - U_1$$

From equation (5.7) change in internal energy

$$= U_2 - U_1 = mc_\mathrm{v}(T_2 - T_1)$$

Therefore, substituting values gives

$$mc_\mathrm{p}(T_2 - T_1) - \frac{P(V_2 - V_1)}{J} = mc_\mathrm{v}(T_2 - T_1)$$

By applying the characteristic gas equation $PV = mRT$ at 1 and 2
$PV_1 = mRT_1$ and $PV_2 = mRT_2$

$$\therefore \qquad \frac{PV_2 - PV_1}{J} = \frac{P(V_2 - V_1)}{J} = \frac{mR(T_2 - T_1)}{J} \qquad (5.9)$$

Substituting in $Q - W/J = U_2 - U_1$ gives

$$mc_p(T_2 - T_1) - \frac{mR(T_2 - T_1)}{J} = mc_v(T_2 - T_1)$$

$$\therefore \qquad\qquad\qquad c_p - \frac{R}{J} = c_v$$

$$\text{or} \qquad\qquad\qquad c_p - c_v = \frac{R}{J} \qquad\qquad (5.10)$$

Since R and J are positive quantities then $c_p - c_v$ must always be a positive quantity, i.e. $c_p > c_v$.

In the chapter on non-flow processes, it was shown that during a constant pressure process, the heat energy supplied is equal to the change in enthalpy. Applying this to a perfect gas gives

$$Q = mc_p(T_2 - T_1) = H_2 - H_1 \qquad (5.11)$$

or alternatively since

$$H_1 = U_1 + \frac{P_1V_1}{J}, \text{ and } H_2 = U_2 + \frac{P_2V_2}{J}$$

$$\text{then} \qquad H_2 - H_1 = (U_2 - U_1) + \frac{(P_2V_2 - P_1V_1)}{J}$$

$$= mc_v(T_2 - T_1) + \frac{mR}{J}(T_2 - T_1)$$

$$= mc_v(T_2 - T_1) + m(c_p - c_v)(T_2 - T_1)$$

$$= m(T_2 - T_1)(c_v + c_p - c_v)$$

$$= mc_p(T_2 - T_1) \qquad\qquad (5.12)$$

Ex. 5.2. A non-conducting cylinder of 4 in bore contains air and is fitted with a frictionless non-conducting piston which weighs 15 lbf. Initially the temperature of the air is 60°F, and the piston is at a height of 6 in above the cylinder base. Heat energy is supplied to the air until the piston rises a further 6 in. Given that atmospheric pressure is 14·7 lbf/in² determine the quantity of heat energy supplied to the air, and the change in internal energy of the air. For air, $c_p = 0.24$ Btu/lb deg R, and $R = 53.3$ ft lbf/lb degR.

Since weight of piston is constant, the process will be at constant pressure conditions.

$$\text{Area of cylinder} = \frac{\pi}{4} \times 4^2 = 12 \cdot 57 \text{ in}^2$$

$\therefore$ initial volume of gas $V_1 = 12 \cdot 57 \text{ in}^2 \times 6 \text{ in}$
$$= 75 \cdot 42 \text{ in}^3 = 0 \cdot 0436 \text{ ft}^3$$

final volume of gas $V_2 = 12 \cdot 57 \text{ in}^2 \times 12 \text{ in} = 0 \cdot 0872 \text{ ft}^3$

$$\text{absolute pressure of gas} = 14 \cdot 7 \text{ lbf/in}^2 + \frac{15}{12 \cdot 57} \text{ lbf/in}^2$$

$$= 15 \cdot 89 \text{ lbf/in}^2$$

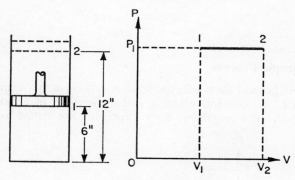

FIGURE E.5.2

From equation (5.8), heat energy supplied $Q = m \times c_p(T_2 - T_1)$

where $m = P_1 V_1 / R T_1$ (from equation (5.4))

$$= \frac{(15 \cdot 89 \times 144)(\text{lbf/ft}^2) \times 0 \cdot 0436 \text{ ft}^3}{53 \cdot 3 \text{ (ft lbf/lb degR)} \times (460 + 60) \text{ degR}}$$

$$= 0 \cdot 0036 \text{ lb}$$

$$T_2 = \frac{P_2 V_2}{P_1 V_1} \times T_1 \text{ (from equation (5.3))}$$

$$= \frac{V_2}{V_1} \times T_1 \text{ since } P_1 = P_2$$

$$= \frac{0 \cdot 0872 \text{ ft}^3}{0 \cdot 0436 \text{ ft}^3} \times (460 + 60) \,^\circ\text{R} = 1{,}040 \,^\circ\text{R}$$

$\therefore Q = m \times C_p \times (T_2 - T_1)$

$$= 0 \cdot 0036 \text{ lb} \times 0 \cdot 24 \frac{\text{Btu}}{\text{lb degR}} \times [1{,}040 - (460 + 60)] \text{ degR}$$

$$= 0 \cdot 45 \text{ Btu}$$

4

Non-flow energy equation gives

$$Q = (U_2 - U_1) + \frac{W}{J}$$

where $\frac{W}{J} = \frac{P(V_2 - V_1)}{J}$

$$= \frac{(15{\cdot}89 \times 144)(\text{lbf/ft}^2) \times (0{\cdot}0872 - 0{\cdot}0436)\text{ft}^3}{778 \text{ ft lbf/Btu}}$$

$$= 0{\cdot}128 \text{ Btu}$$

$\therefore$ change in internal energy $= (U_2 - U_1) = Q - \dfrac{W}{J}$

$$= 0{\cdot}45 - 0{\cdot}128$$
$$= 0{\cdot}322 \text{ Btu}$$

5.7 Polytropic Process

The most general form of expressing a process is by means of the equation $PV^n = $ constant, where n, called the index of expansion (or compression), is a number. Such processes are called polytropic

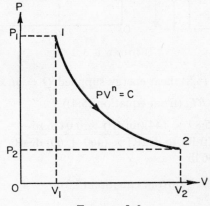

FIGURE 5.6

processes. In such a process both work and heat energy may be transferred across the boundary. Consider m lb of gas contained in a cylinder at conditions P_1, V_1, and T_1. Let the gas undergo a reversible polytropic process of the form $PV^n = c$ to the conditions P_2, V_2, and T_2.

Since $PV^n = $ constant, then

$$P_1 V_1{}^n = P_2 V_2{}^n \tag{5.13}$$

also since $PV/T = $ constant then

$$\frac{P_1 V_1}{T_1} = \frac{P_2 V_2}{T_2} \qquad (5.14)$$

Dividing (5.13) by (5.14) gives

$$\frac{P_1 V_1{}^n}{P_1 V_1/T_1} = \frac{P_2 V_2{}^n}{P_2 V_2/T_2}$$

Therefore

$$\frac{T_1 V_1{}^n}{V_1} = \frac{T_2 V_2{}^n}{V_2}$$

or

$$T_1 V_1{}^{n-1} = T_2 V_2{}^{n-1}$$

∴

$$\frac{T_2}{T_1} = \left(\frac{V_1}{V_2}\right)^{n-1} \qquad (5.15)$$

or

$$\frac{V_1}{V_2} = \left(\frac{T_2}{T_1}\right)^{1/(n-1)} \qquad (5.16)$$

also

$$\left(\frac{V_1}{V_2}\right)^n = \frac{P_2}{P_1}$$

∴

$$\frac{V_1}{V_2} = \left(\frac{P_2}{P_1}\right)^{1/n}$$

∴

$$\left(\frac{V_1}{V_2}\right)^{n-1} = \frac{T_2}{T_1} = \left[\left(\frac{P_2}{P_1}\right)^{1/n}\right]^{n-1} = \left(\frac{P_2}{P_1}\right)^{(n-1)/n}$$

or

$$\frac{P_2}{P_1} = \left(\frac{T_2}{T_1}\right)^{n/(n-1)} \qquad (5.17)$$

From §5.5 it is seen that the change in internal energy during the process $= U_2 - U_1 = mc_v(T_2 - T_1)$. Also, the work energy transferred during the process:

$$\frac{W}{J} = \frac{P_1 V_1 - P_2 V_2}{J(n-1)}$$

heat units (see Appendix 2.1).

Applying the characteristic gas equation at 1 and 2 gives

$$P_1 V_1 = mRT_1 \text{ and } P_2 V_2 = mRT_2$$

Substituting for $P_1 V_1$ and $P_2 V_2$ gives

$$\frac{W}{J} = \frac{mRT_1 - mRT_2}{J(n-1)} = \frac{mR(T_1 - T_2)}{J(n-1)}$$

Applying non-flow energy equation between 1 and 2 gives

$$Q - W/J = U_2 - U_1$$

Substituting values from above gives

$$Q - \frac{mR(T_1 - T_2)}{J(n - 1)} = mc_{\mathrm{v}}(T_2 - T_1)$$

$$\therefore \qquad Q = \frac{mR(T_1 - T_2)}{J(n - 1)} + mc_{\mathrm{v}}(T_2 - T_1)$$

$$= m(T_1 - T_2)\left(\frac{R}{J(n - 1)} - c_{\mathrm{v}}\right) \qquad (5.18)$$

Since $R/J = c_{\mathrm{p}} - c_{\mathrm{v}}$ from equation (5.10) then

$$Q = m(T_1 - T_2)\left(\frac{c_{\mathrm{p}} - c_{\mathrm{v}}}{n - 1} - c_{\mathrm{v}}\right)$$

$$= m(T_1 - T_2)\left(\frac{c_{\mathrm{p}} - c_{\mathrm{v}} - nc_{\mathrm{v}} + c_{\mathrm{v}}}{n - 1}\right)$$

$$= m(T_1 - T_2)\left(\frac{c_{\mathrm{p}} - nc_{\mathrm{v}}}{n - 1}\right) \qquad (5.19)$$

Ex. 5.3. A cylinder contains 3 ft³ of a gas at 15 lbf/in² and 100°F. The gas is compressed according to the law $PV^{1\cdot3} =$ constant until the pressure is 80 lbf/in². Determine the heat energy supplied or rejected during the process. For the gas take $C_{\mathrm{v}} = 0\cdot17$ Btu/lb degR and $R = 53\cdot3$ ft lbf/lb degR.

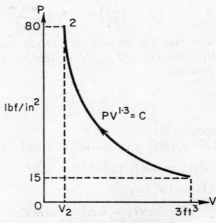

FIGURE E.5.3

The non-flow energy equation gives

$$\text{heat energy supplied } Q = (U_2 - U_1) + \frac{W}{J}$$

For a perfect gas $U_2 - U_1 = m \times c_{\mathrm{v}} \times (T_2 - T_1)$ $\qquad$ (§5.5)
For a process following the law $PV^n = \text{constant}$

$$\frac{W}{J} = \frac{(P_1 V_1 - P_2 V_2)}{J(n-1)}$$

For a perfect gas $P_1 V_1 = mRT_1$ and $P_2 V_2 = mRT_2$

$$\therefore \quad \frac{W}{J} = \frac{(mRT_1 - mRT_2)}{J(n-1)} = \frac{mR}{J(n-1)}(T_1 - T_2)$$

From equation (5.17)

$$\frac{T_2}{T_1} = \left(\frac{P_2}{P_1}\right)^{(n-1)/n}$$

$$\therefore \quad \frac{T_2}{(460 + 100)\,^\circ R} = \left(\frac{80\ \mathrm{lbf/in^2}}{15\ \mathrm{lbf/in^2}}\right)^{(1\cdot3-1)/1\cdot3}$$
$$= (5\cdot33)^{0\cdot231}$$
$$= 1\cdot472$$
$$\therefore \quad T_2 = 1\cdot472 \times 560\,^\circ R = 825\,^\circ R$$

also
$$m = \frac{P_1 V_1}{RT_1}$$
$$= \frac{(15 \times 144)\,(\mathrm{lbf/ft^2}) \times 3\ \mathrm{ft^3}}{53\cdot3(\mathrm{ft\ lbf/lb\ degR}) \times (460 + 100)\,^\circ R}$$
$$= 0\cdot217\ \mathrm{lb}$$

$$\therefore \quad U_2 - U_1 = m \times c_{\mathrm{v}} \times (T_2 - T_1)$$
$$= 0\cdot217\ \mathrm{lb} \times 0\cdot17\ \frac{\mathrm{Btu}}{\mathrm{lb\ degR}} \times (825 - 560)\ \mathrm{degR}$$
$$= 9\cdot77\ \mathrm{Btu}$$
$$\frac{W}{J} = \frac{mR}{J(n-1)}(T_1 - T_2)$$
$$= \frac{0\cdot217\ \mathrm{lb} \times 53\cdot3\ \mathrm{ft\ lbf/lb\ degR}}{778\ (\mathrm{ft\ lbf/lb\ degR}) \times (1\cdot3 - 1)}(560 - 825)\ \mathrm{degR}$$
$$= -13\cdot12\ \mathrm{Btu}$$
$$\therefore \quad Q = (U_2 - U_1) + \frac{W}{J}$$
$$= 9\cdot77\ \mathrm{Btu} + (-13\cdot12)\ \mathrm{Btu}$$
$$= -3\cdot35\ \mathrm{Btu}$$

i.e. 3·35 Btu are rejected during the process.

5.8 Reversible Adiabatic Process

In the special case of a reversible process where no heat energy is transferred to or from the gas the process will be a reversible adiabatic process. These special processes are called <u>isentropic</u> processes. Equation (5.19) will then become

$$Q = m(T_1 - T_2)\frac{(c_p - nc_v)}{n - 1} = 0$$

$$\therefore \qquad c_p - nc_v = 0$$

$$\therefore \qquad \frac{c_p}{c_v} = n$$

It must be emphasised that this last equation is only true in the case of a reversible adiabatic process. The ratio c_p/c_v is called the isentropic index and is denoted by γ. Thus, only in the special case where a reversible polytropic process is adiabatic, may the index n be replaced by γ, where

$$\gamma = \frac{c_p}{c_v} \qquad (5.20)$$

Therefore for a reversible adiabatic process the equation of the process becomes $PV^\gamma = $ constant and the expression for W/J becomes

$$\frac{W}{J} = \frac{P_1V_1 - P_2V_2}{J(\gamma - 1)} = \frac{mR(T_1 - T_2)}{J(\gamma - 1)}$$

Similarly

$$\frac{T_2}{T_1} = \left(\frac{V_1}{V_2}\right)^{\gamma-1} \text{ or } \frac{V_1}{V_2} = \left(\frac{T_2}{T_1}\right)^{1/\gamma-1} \qquad (5.21)$$

and

$$\frac{T_2}{T_1} = \left(\frac{P_2}{P_1}\right)^{(\gamma-1)/\gamma} \text{ or } \frac{P_2}{P_1} = \left(\frac{T_2}{T_1}\right)^{\gamma/(\gamma-1)} \qquad (5.22)$$

Ex. 5.4. 0·5 lb of air at 120 lbf/in² and 1,000°F expands adiabatically and reversibly to a temperature of 300°F. Determine the final pressure, final volume, and the work energy transferred during the process. For air, $c_p = 0·24$ Btu/lb degR, and $R = 53·3$ ft lbf/lb degR.

For a reversible adiabatic process for a perfect gas, $PV^\gamma = $ constant. From equation (5.22)

$$\frac{P_2}{P_1} = \left(\frac{T_2}{T_1}\right)^{\gamma/(\gamma-1)}$$

$$\therefore \quad \frac{P_2}{120 \text{ lbf/in}^2} = \left[\frac{(460 + 300\,°R)}{(460 + 1,000\,°R)}\right]^{\gamma/(\gamma-1)} = \left(\frac{760}{1,460}\right)^{\gamma/(\gamma-1)}$$

Now

$$\gamma = \frac{c_p}{c_v}$$

and

$$c_p - c_v = \frac{R}{J}$$

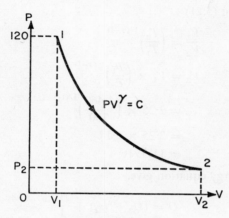

FIGURE E.5.4

$$\therefore \quad c_v = c_p - \frac{R}{J}$$

$$= 0.24 \ (\text{Btu/lb degR}) - \frac{53.3 \ \text{ft lbf/lb degR}}{778 \ \text{ft lbf/Btu}}$$

$$= 0.24 - 0.0685$$

$$= 0.1715 \ \text{Btu/lb degR}$$

$$\therefore \quad \gamma = \frac{c_p}{c_v} = \frac{0.24 \ \text{Btu/lb degR}}{0.1715 \ \text{Btu/lb degR}} = 1.4$$

From above

$$P_2 = 120 \ (\text{lbf/in}^2) \times \left(\frac{760}{1,460}\right)^{1.4/(1.4-1)}$$

$$= 120 \ (\text{lbf/in}^2) \times (0.52)^{3.5}$$

$$= 120 \ (\text{lbf/in}^2) \times 0.101$$

$$= 12.1 \ \text{lbf/in}^2$$

The characteristic gas equation gives $PV = mRT$

Applied at 1 this gives

$$V_1 = \frac{mRT_1}{P_1}$$

$$= \frac{0{\cdot}5 \text{ lb} \times 53{\cdot}3 \text{ (ft lbf/lb degR)} \times (460 + 1{,}000) \text{ degR}}{(120 \times 144) \text{ lbf/ft}^2}$$

$$= 2{\cdot}25 \text{ ft}^3$$

From equation (5.21)

$$\therefore \qquad \frac{V_1}{V_2} = \left(\frac{T_2}{T_1}\right)^{1/(\gamma-1)}$$

$$V_2 = V_1 \times \left(\frac{T_1}{T_2}\right)^{1/(\gamma-1)}$$

$$= 2{\cdot}25 \times \left(\frac{1{,}460}{760}\right)^{1/(1{\cdot}4-1)}$$

$$= 2{\cdot}25 \times 5{\cdot}14$$

$$= 11{\cdot}55 \text{ ft}^3$$

Non-flow energy equation gives

$$Q = (U_2 - U_1) + \frac{W}{J}$$

For an adiabatic process

$$Q = 0$$

$$\therefore \frac{W}{J} = -(U_2 - U_1)$$

$$= -m \times c_\text{v} \times (T_2 - T_1)$$

$$= -0{\cdot}5 \text{ lb} \times 0{\cdot}1715 \frac{\text{Btu}}{\text{lb degR}} \times (760 - 1{,}460) \text{ degR}$$

$$= 60 \text{ Btu}$$

5.9 Isothermal Process

The characteristic gas equation gives $PV = mRT$. If, during the process, T remains constant (isothermal process) then this equation becomes $PV =$ constant, which may be represented by the curve shown in Fig. 5.7.

Consider m lb of a gas contained in a cylinder at pressure P_1, volume V_1, and temperature T_1. Let the gas undergo an isothermal process to pressure P_2 and volume V_2 (i.e. state 2). Since the volume changes a transfer of work energy will take place. As shown in the Appendix 2 if a process follows the equation $PV = $ constant, then the expression for the work energy transfer is given by

$$\frac{W}{J} = \frac{P_1 V_1}{J} \ln \left(\frac{V_2}{V_1} \right) \text{ heat units} \qquad (5.23)$$

Since there is no change in temperature, then by Joule's Law there will be no change in internal energy. Thus the non-flow energy equation

$$Q - W/J = U_2 - U_1$$

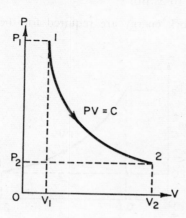

FIGURE 5.7

will become

$$Q - W/J = 0$$

or

$$Q = W/J$$

∴

$$Q = \frac{W}{J} = \frac{P_1 V_1}{J} \ln \left(\frac{V_2}{V_1} \right) \qquad (5.24)$$

During the isothermal expansion of a gas, when W is positive, then Q must also be positive, i.e. heat energy must be supplied to the gas as it expands, if the temperature of the gas is to be maintained constant.

Ex. 5.5. 2 ft³ of air at 20 lbf/in² are compressed isothermally to a volume of 0·5 ft³. Determine the work energy required for the compression and compare this with the work energy which would be required for a reversible adiabatic compression through the same volume ratio. For air, $R = 53\cdot3$ ft lbf/lb degR and $\gamma = 1\cdot4$.

For an isothermal process for a perfect gas, $PV = $ constant.

i.e. $$P_1V_1 = P_2V_2$$

and $$\frac{W}{J} = \frac{P_1V_1}{J} \ln \frac{V_2}{V_1} \qquad \text{(equation (5.23))}$$

$$= \frac{(20 \times 144)\,(\text{lbf/ft}^2) \times 2\,\text{ft}^3}{778\,\text{ft lbf/Btu}} \ln \left(\frac{0 \cdot 5\,\text{ft}^3}{2\,\text{ft}^3} \right)$$

$$= 7 \cdot 4 \times (-1 \cdot 388)$$

$$= -10 \cdot 28\,\text{Btu}$$

i.e. 10·28 Btu of work energy are required for the isothermal compression.

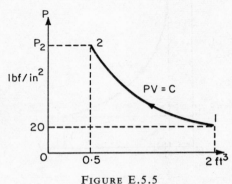

FIGURE E.5.5

If the compression process was a reversible adiabatic, $PV^\gamma = $ constant.

$$P_1V_1{}^\gamma = P_xV_x{}^\gamma$$

$$\therefore \qquad P_x = P_1 \left(\frac{V_1}{V_x} \right)^\gamma$$

$$= 20\,(\text{lbf/in}^2) \times \left(\frac{2\,\text{ft}^3}{0 \cdot 5\,\text{ft}^3} \right)^{1 \cdot 4}$$

$$= 20\,(\text{lbf/in}^2) \times 6 \cdot 98$$

$$= 139 \cdot 6\,\text{lbf/in}^2$$

For a reversible adiabatic process

$$\frac{W}{J} = \frac{P_1V_1 - P_2V_2}{J(\gamma - 1)}$$

$$= \frac{(20 \times 144)\,(\text{lbf/ft}^2) \times 2\,\text{ft}^3 - (139 \cdot 6 \times 144)(\text{lbf/ft}^2) \times 0 \cdot 5\,\text{ft}^3}{778\,(\text{ft lbf/Btu}) \times (1 \cdot 4 - 1)}$$

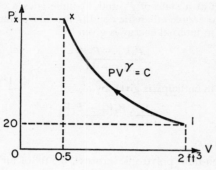

FIGURE E.5.6

$$= \frac{5,760 \text{ ft lbf} - 10,050 \text{ ft lbf}}{778 \text{ (ft lbf/Btu)} \times 0 \cdot 4}$$

$$= -\frac{5 \cdot 52 \text{ Btu}}{0 \cdot 4}$$

$$= -13 \cdot 8 \text{ Btu}$$

i.e. 13·8 Btu of work energy are required for adiabatic compression.

EXERCISES ON CHAPTER 5

N.B. All pressures are absolute.

1. A cylinder contains 10 ft³ of a perfect gas at 15 lbf/in² and 85°F. The gas is compressed according to the law $PV^{1 \cdot 3} = $ constant until the volume is reduced to 1 ft³. Heat energy is then supplied at constant pressure until the volume becomes 2 ft³. Determine:

(a) The temperature and pressure at the end of each process
(b) The total change in internal energy
(c) The work energy transfer during each process

For the gas take $c_p = 0 \cdot 255$ Btu/lb degF, $c_v = 0 \cdot 18$ Btu/lb degF.

(1,088°R, 300 lbf/in²; 2,176°R, 300 lbf/in²; +199 Btu; −92·5 Btu; +55·5 Btu)

2. During a non-flow process, 1·5 ft³ of gas at 120 lbf/in² and 600°F are to be expanded to a volume of 6 ft³. If $c_p = 0 \cdot 24$ Btu/lb degF and $c_v = 0 \cdot 17$ Btu/lb degF, calculate the work energy transfer and the final temperature if the process is (a) hyperbolic, (b) adiabatic.

(+46·2 Btu, 1,060°R; +35·3 Btu, 600°R)

3. A quantity of gas at 250 lbf/in² and 300°F undergoes a non-flow constant pressure process which causes the volume of the gas to increase from 1 ft³ to 5 ft³. If $c_p = 0 \cdot 24$ Btu/lb degF and $c_v = 0 \cdot 171$ Btu/lb degF, determine:

(a) The change in internal energy
(b) The work energy transferred
(c) The heat energy transferred

(+458 Btu; +185 Btu; +643 Btu)

4. A perfect gas at a state of P_1 and V_1 undergoes a change of state to P_2 and V_2 under reversible adiabatic conditions. Show that

    (*a*) The change in internal energy is given by

$$\frac{P_2V_2 - P_1V_1}{J(\gamma - 1)}$$

    (*b*) The change in enthalpy is given by

$$\frac{\gamma}{\gamma - 1}\frac{P_2V_2 - P_1V_1}{J}$$

where $\gamma = c_p/c_v$

5. During a non-flow polytropic process, 6.5 ft^3 of air at 15 lbf/in^2 and $80°$F is compressed until the pressure and temperature become 180 lbf/in^2 and $540°$F respectively. If $c_p = 0.238$ Btu/lb degF, and $c_v = 0.17$ Btu/lb degF, determine

    (*a*) The value of the index of compression
    (*b*) The work energy transfer
    (*c*) The change in internal energy
    (*d*) The heat energy transfer

                              (1.33; -46.7 Btu; $+38.4$ Btu; -8.3 Btu)

6. A quantity of gas occupying 0.5 ft^3 at 140 lbf/in^2 and $700°$F is heated during a constant volume process until the pressure reaches 600 lbf/in^2. The gas is then expanded adiabatically to a pressure of 40 lbf/in^2. Given that $R = 53.5$ ft lbf/lb degF and $\gamma = 1.41$, calculate:

    (*a*) The temperature at the beginning of expansion
    (*b*) The temperature at the end of expansion
    (*c*) The heat energy supplied
    (*d*) The work energy transferred

                      ($4,970°$R; $2,260°$R; 104 Btu; $+73.8$ Btu)

7. Show that during the expansion of a quantity of gas according to the law PV^n = constant, the heat energy transfer Q and the work energy transfer W are connected by the expression

$$Q = \frac{\gamma - n}{\gamma - 1}\frac{W}{J}$$

8. Gas at 15 lbf/in^2 and $70°$F; with a specific volume of 12 ft^3/lb undergoes adiabatic compression through a volume ratio of 6:1. During the compression the work energy transfer is 7 Btu, and the final temperature of the gas is $600°$F. Calculate the value of γ, R, C_p, C_v, and also the mass of gas involved in the process.

   (1.387; 48.9 ft lbf/lb deg F; 0.225 Btu/lb deg F; 0.164 Btu/lb deg F; 0.0813 lb)

9. 5 ft^3 of gas at 20 lbf/in^2 and $100°$F are compressed during a non-flow process according to the law $PV^{1.35}$ = constant, to a pressure of 300 lbf/in^2. If $R = 49$ ft lbf/lb deg F, and $\gamma = 1.41$ determine the change in internal energy, the work energy transfer, and the heat energy transfer.

                          ($+44.8$ Btu; -53.7 Btu; -8.9 Btu)

6: The Second Law of Thermodynamics

The First Law gave an exact equivalence of heat energy and work. There is no guarantee, however, that a cycle will take place even though it fulfils the First Law. Every cycle must obey the First Law but in addition another law must be satisfied. It is common knowledge that all rivers flow downhill and that bodies at different temperatures tend to a common temperature but the reverse of these simple processes does not occur in practice. In thermodynamics the existence of a similar directional effect, namely that heat energy flows from a hot to a cold body gives rise to the Second Law of Thermodynamics. An important aspect of the Second Law is its application to heat engine cycles.

6.1 Heat Engine Cycle

A heat engine cycle consists of a number of processes arranged to convert heat energy into work energy such that the system is returned to its original state at the end of each cycle. Consider a heat engine arranged as in Fig. 6.1.

Suppose that this engine is arranged to lift a load w from platform A to platform B. If heat energy Q_s is supplied to the system the piston moves upwards from A to B and an amount of work W_1 is done in lifting the load w, the piston, and its attachments. Let the load now be removed to platform B. If heat energy Q_r is now rejected from the system such that the piston is returned to its original position an amount of work W_2 will be done on the system by the weight of the piston and its attachment in falling from B to A. The system, which has now returned to its original state, has undergone a closed cycle and, therefore, its change of internal energy will be zero.

The net work done

$$W_n = W_1 - W_2$$

or, in heat units

$$\frac{W_n}{J} = \frac{W_1}{J} - \frac{W_2}{J} \qquad (6.1)$$

The net heat transfer

$$Q_n = Q_s - Q_r$$

The First Law requires that

$$\frac{W_n}{J} = Q_n$$

or
$$\frac{W_n}{J} = Q_s - Q_r \qquad (6.2)$$

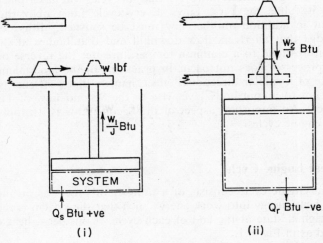

FIGURE 6.1

The efficiency of the engine, when measured as its effectiveness in transforming the heat energy supplied into work energy is given by

$$\text{efficiency} = \frac{\text{net work done}}{\text{heat energy supplied}}$$

i.e.
$$= \frac{W_n/J}{Q_s}$$

$$= \frac{Q_s - Q_r}{Q_s}$$

$$= 1 - \frac{Q_r}{Q_s} \qquad (6.3)$$

It is obvious from this study that a certain amount of heat energy Q_r had to be rejected from the system in order to return the piston to its original state, and also that the efficiency can only be unity or 100% if Q_r is zero. Since Q_r has a finite value, then the efficiency must be less than 100%. This result means that it is not possible to convert all the heat energy available into work and points to the existence of the Second Law. A heat engine cycle may be illustrated simply as shown in Fig. 6.2. In, for instance, a steam turbine plant the source is the combustion chamber of the boiler and the sink is the cooling water in the condenser.

Since the net work is never equal to Q_s there is always some heat

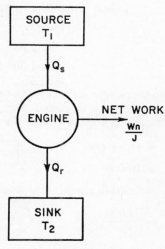

FIGURE 6.2

energy rejected to the sink. It is also known from experience, that there will never be any work done unless there is a temperature difference between the source and the sink. The recognition of these simple facts was first reported by Sadi Carnot in 1824. The Second Law, which is a statement of these facts, was given in the form: 'Wherever a temperature difference exists, motive power can be produced'. The law may be stated in many forms each revealing a different viewpoint of the fundamental truth. It is often quoted in the form: 'It is impossible for a heat engine to produce work if it exchanges heat with a source and with a sink which are at the same temperature'. Although there is no mathematical proof of the Second Law, no one has yet been able to describe or construct an engine which violates the Law. If such an engine were possible it would be called a 'perpetual motion machine of the second kind'. A perpetual motion machine of the first kind would be one which violated the First Law.

Ex. 6.1. An engine receives heat at the rate of 2,000 Btu/min and produces 15 h.p. at the output shaft. Determine the efficiency of the engine and the heat rejected per minute.

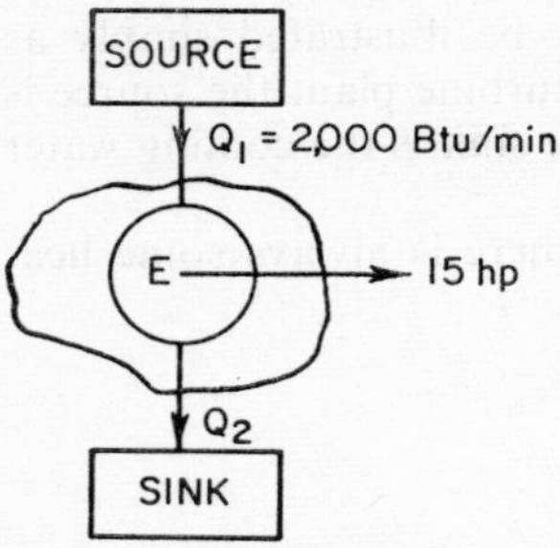

FIGURE E.6.1

$$15 \text{ h.p.} = \frac{15 \times 33,000}{778}$$

$$= 636 \text{ Btu/min}$$

$$\text{efficiency} = \frac{636}{2,000} \times 100\% = 31 \cdot 8\%$$

Heat rejected per min

$$Q_2 = Q_1 - W$$
$$= 2,000 - 636$$
$$= 1,364 \text{ Btu/min}$$

Ex. 6.2. A heat pump supplies heat energy to a building at the rate of 10,000 Btu/min. The heat energy extracted from the cold source is 8,000 Btu/min. Find the input horsepower to the pump.

$$W = Q_1 - Q_2$$
$$= 10,000 - 8,000$$
$$= 2,000 \text{ Btu/min}$$

$$1 \text{ h.p.} = \frac{33,000}{778}$$

$$= 42 \cdot 4 \text{ Btu/min}$$

$$\therefore \quad \text{h.p. of pump} = \frac{2,000}{42 \cdot 4} = 47 \cdot 4$$

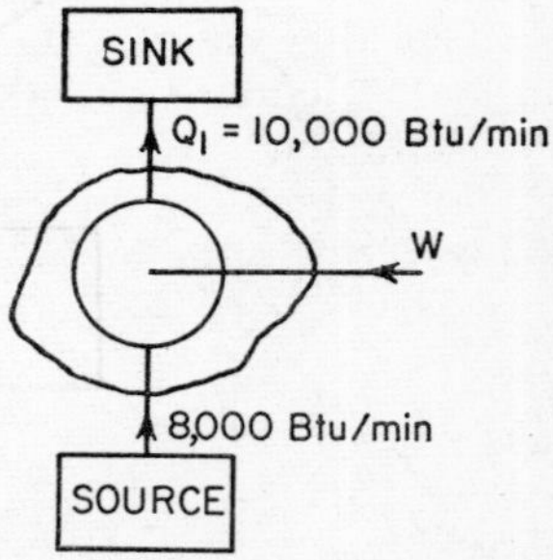

FIGURE E.6.2

6.2 Reversibility

From what has been stated it can be seen that heat energy is transferred naturally from a higher to a lower temperature—a directional effect. In this case a process undergoes a change in a certain direction. If this process can be reversed so that the path which was traced can be exactly retraced and the system and its surroundings restored to the original state, then the process is said to be *reversible*. No evidence would exist that the process had ever taken place. In practice, no such process is possible, but the concept of reversibility is, nevertheless, extremely useful. A pendulum swinging in a vacuum on a sharp knife edge illustrates a process which is almost reversible. The pendulum starts from one extremity of its swing and at the bottom of the swing its potential energy has been converted into kinetic energy. This, in turn, is reconverted into potential energy at the other extremity of the swing. The pendulum then retraces its path, and very nearly reaches the point from which it started, but not quite, for some of the energy has been used to overcome the friction at the knife edge. It is seen therefore, that the presence of friction renders a process *irreversible*.

Consideration of the Second Law has shown that any one of the following conditions will prevent a process from being reversible:

(*a*) The presence of friction
(*b*) Heat energy transfer from a higher to a lower temperature
(*c*) Unresisted expansion (or free expansion) from a higher to a lower pressure
(*d*) Paddle work
(*e*) Lack of pressure equilibrium throughout the system

It is obvious that a reversible process is an ideal which is impossible in reality, and it can be stated briefly that in order for a process to be reversible the following conditions must always apply:

1. The process may proceed in either direction
2. That the system is always in a state of equilibrium
3. That friction is absent
4. That energy transformations which take place whilst the process proceeds in one direction must equal those which take place as it proceeds in the opposite direction
5. That the system and its surroundings can be restored to its original condition by retracing in the reverse direction every point it took in the original direction

As previously indicated in chapter 1, only processes which are a succession of equilibrium states (reversible processes) can be represented on P–V or other state diagrams. Only the initial and final states are known in irreversible processes.

6.3 Carnot Cycle

Sadi Carnot proposed a heat engine cycle composed of reversible processes which gave the maximum possible work when working between two fixed temperature limits, T_1 and T_2. Consider an engine in which a frictionless weightless piston is fitted in a cylinder (Fig. 6.3). Suppose that heat energy Q_1 is made to flow into an ideal fluid enclosed in the cylinder at a constant temperature (i.e. an isothermal process) so that the fluid expands from a to b. Imagine now an adiabatic expansion from b to c, followed by an isothermal compression at temperature T_2 from c to d during which heat Q_2 is rejected to a sink.

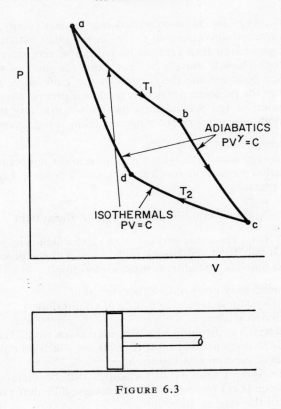

FIGURE 6.3

Finally, an adiabatic compression from d to a closes the cycle. All of these processes are reversible.

The net work done = heat supplied − heat rejected

$$Q_1 - Q_2$$

and the efficiency $= \dfrac{Q_1 - Q_2}{Q_1}$

$$= 1 - \frac{Q_2}{Q_1} \tag{6.4}$$

It can be shown that this efficiency may be expressed as $1 - (T_2/T_1)$ as follows:

Assume that the fluid is a perfect gas. Firstly, consider the isothermal from a to b.

Applying

$$Q = (U_2 - U_1) + \frac{W}{J}$$

it is seen that, since the temperature does not change,

$$U_2 - U_1 = mc_v(T_2 - T_1) = 0 \quad \text{since } T_2 = T_1$$

and therefore

$$Q_1 = \frac{W}{J} = \frac{P_a V_a}{J} \ln \frac{V_b}{V_a}$$

$$= \frac{mRT_1}{J} \ln \frac{V_b}{V_a} \tag{6.5}$$

Consider, now, the adiabatic expansion from b to c. Here, $Q = 0$ and therefore the

work done by the gas = change of internal energy
$$= mc_v(T_1 - T_2)$$

It is evident that the temperature must fall to T_2 in order that work may be done.

During the isothermal compression from c to d there is no change of internal energy and, therefore, the heat rejected

$$Q_2 = \frac{W}{J}$$

$$= \frac{P_c V_c}{J} \ln \frac{V_c}{V_d}$$

$$= \frac{mRT_2}{J} \ln \frac{V_c}{V_d} \tag{6.6}$$

In the adiabatic compression from d to a, the heat supplied = 0, and the work done is equal to the gain in internal energy = $mc_v(T_1 - T_2)$. The net work done = $Q_1 - Q_2$

$$= mRT_1 \ln \frac{V_b}{V_a} - mRT_2 \ln \frac{V_c}{V_d}$$

and the efficiency = $1 - \dfrac{Q_2}{Q_1}$

$$= 1 - \frac{mRT_2 \ln (V_c/V_d)}{mRT_1 \ln (V_b/V_a)}$$

$$= 1 - \frac{T_2 \ln (V_c/V_d)}{T_1 \ln (V_b/V_a)} \tag{6.7}$$

For the adiabatic expansion, b to c

$$\frac{T_1}{T_2} = \left(\frac{V_c}{V_b}\right)^{\gamma-1} \tag{6.8a}$$

where $PV^\gamma = C$ is the equation of the expansion and compression. For the compression, d to a

$$\frac{T_1}{T_2} = \left(\frac{V_d}{V_a}\right)^{\gamma-1} \tag{6.8b}$$

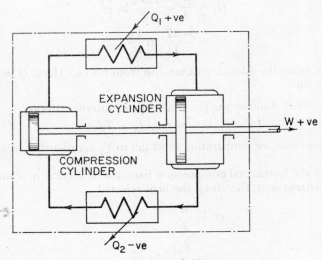

FIGURE 6.4

From equations (6.8a) and (6.8b) is obtained

$$\frac{V_c}{V_b} = \frac{V_d}{V_a}$$

or

$$\frac{V_c}{V_d} = \frac{V_b}{V_a} \tag{6.9}$$

Substituting (6.9) in equation (6.7) gives the Carnot efficiency

$$\eta = 1 - \frac{T_2}{T_1} = \frac{T_1 - T_2}{T_1} \tag{6.10}$$

It is important to note that these temperatures are absolute values and that the value of the efficiency increases as T_1 increases or as T_2 decreases.

The limit imposed upon T_1 in practice is the metallurgical consideration of the metals used in the engine. The expression for the Carnot efficiency is independent of the properties of the fluid so that the expression holds for any fluid whether it be a gas or a vapour.

An engine working under steady flow conditions on the Carnot cycle can be represented by Fig. 6.4.

Ex. 6.3. A steam turbine plant works between temperature limits of 1,000°F and 110°F. Determine the Carnot efficiency of the plant.

$$T_1 = 1,000 + 460 = 1,460°R$$

$$T_2 = 110 + 460 = 570°R$$

$$\text{Carnot efficiency} = \frac{T_1 - T_2}{T_1}$$

$$= \frac{1,460 - 570}{1,460}$$

$$= \frac{890}{1,460}$$

$$= 0.61 \text{ or } 61\%$$

6.4 Refrigerators

The reversed Carnot cycle as a refrigerator is shown in Fig. 6.5. Here heat Q_2 is extracted from the cold chamber and heat Q_1 is rejected

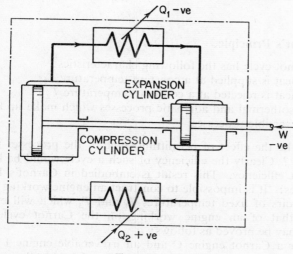

FIGURE 6.5

into the surroundings. W/J is the net work done. The effectiveness of the refrigerator is measured by

$$\frac{\text{heat extracted } Q_2}{\text{work done } W/\text{J}}$$

This is termed the coefficient of performance and can be proved to be equal to $T_2/(T_1 - T_2)$ in the case of reversed Carnot cycle.

6.5 Heat Pump

The heat pump cycle is precisely similar to Fig. 6.5 but the main consideration of the heat pump is the heat rejected which is used for space heating of buildings. The term performance energy ratio has been used to measure its effectiveness.

$$\text{P.E.R.} = \frac{\text{heat rejected}}{\text{work done}} = \frac{Q_1}{W/\text{J}}$$

$$= \frac{Q_1}{Q_1 - Q_2} = \frac{T_1}{T_1 - T_2}$$

for the reversed Carnot cycle.

The Second Law has also been stated in the following manner:

'It is impossible for heat to be transferred from a lower to a higher temperature without the expenditure of external work'. In this form it is particularly applicable to refrigerators or to heat pumps where it was seen that external work had to be supplied to effect the heat transference.

6.6 Carnot's Principle

The Carnot cycle has the following characteristics:

1. The heat is supplied at a constant temperature T_1
2. The heat is rejected at a constant temperature T_2
3. The isothermal and adiabatic processes which make up the cycle are reversible processes

What will be the effect of permitting some of the processes to occur irreversibly? Clearly the efficiency of such a cycle would be less than the Carnot efficiency. This result is embodied in Carnot's Principle which states: 'It is impossible to construct an engine working between two reservoirs of fixed temperatures T_1 and T_2 which will exceed in efficiency that of an engine working on the Carnot cycle'. This principle may be proved as follows:

Consider a Carnot engine C and an irreversible engine I working between two heat reservoirs as shown in Fig. 6.6. Let the engines be coupled so that the work developed by C is equal to that absorbed by I, when I works as a refrigerator.

Assume that

$$\eta_I > \eta_c$$

Now $\eta_I = \dfrac{W}{Q_x}$ and $\eta_c = \dfrac{W}{Q_1}$

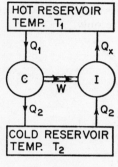

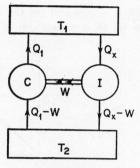

FIGURE 6.6   FIGURE 6.7

According to the assumption

$$\frac{W}{Q_x} > \frac{W}{Q_1}$$

or

$$Q_x < Q_1$$

Now, imagine the engine I to drive the engine C as a refrigerator as shown in Fig. 6.7. The hot reservoir will now gain $Q_1 - Q_x$ heat units and the cold reservoir will lose $(Q_1 - W) - (Q_x - W) = Q_1 - Q_x$ heat units.

Since $Q_x < Q_1$, then this result means that a positive quantity of heat energy $Q_1 - Q_x$ is transferred from cold body to a hot body without the expenditure of any external work. This is contrary to the Second Law and it must, therefore, be concluded that Carnot's principle is correct, and that η_I is less than η_c.

6.7 The Carnot Cycle and the Absolute Temperature Scale

The Carnot cycle efficiency depends upon the temperatures only, and is independent of the properties of the working fluid.

Since the efficiency

$$\eta = 1 - \frac{T_2}{T_1} = 1 - \frac{Q_2}{Q_1} \qquad (6.10)$$

it can be said that

$$\frac{T_2}{T_1} = \frac{Q_2}{Q_1} \qquad (6.11)$$

i.e. Q_2/Q_1 is dependent on the temperatures only or in mathematical language Q_2/Q_1 is a function of temperature only.

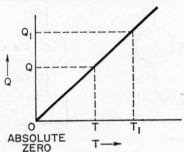

FIGURE 6.8

If T_1 is given a fixed number then

$$T_2 = \frac{Q_2}{Q_1} \times T_1$$

or for any temperature T

$$T = Q \frac{T_1}{Q_1} \qquad (6.12)$$

Equation (6.12) defines a *thermodynamic* scale of temperatures in which T_1 is a fixed temperature and Q and Q_1 are the heats rejected and supplied in an engine working on the Carnot cycle between the temperatures T and T_1. The scale is entirely independent of the fluid properties and is a consequence of the Second Law. A graph of the equation (6.12) is given in Fig. 6.8. It is seen that when $Q = 0$, the temperature T becomes zero. This fixes the absolute zero of temperature.

Imagine now a series of reversible engines arranged in series, each producing the same amount of work. It can be seen from the graph that the temperature drops across the engines will be equal (Fig. 6.9). If enough engines are placed in series so that the total heat transfer $= \Sigma W = Q_1$, then the heat rejected by the last engine will be zero and the temperature at this point will be zero. This last statement gives rise to an interesting point; the efficiency of the last engine will be 100% since it would convert all the heat supplied to it into work. This is an impossibility according to the Second Law and it has been suggested that the interpretation of this apparent contradiction is that it is

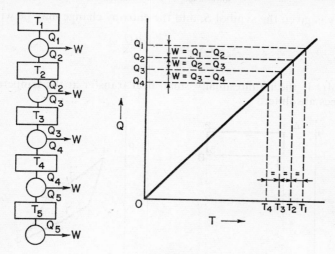

FIGURE 6.9

impossible to reach the absolute zero of temperature and that the scale must stop at an infinitesimal distance from the absolute zero.

6.8 Entropy

From equation (6.11), for the Carnot cycle

$$\frac{Q_1}{T_1} = \frac{Q_2}{T_2}$$

or

$$\frac{Q_1}{T_1} - \frac{Q_2}{T_2} = 0 \tag{6.13}$$

Since Q_2 is negative in itself with respect to Q_1 this equation is sometimes written

$$\frac{Q_1}{T_1} - \frac{(-Q_2)}{T} = \frac{Q_1}{T_1} + \frac{Q_2}{T_2} = 0$$

Both Q_1 and Q_2 are the heat transfers at constant temperatures. The function Q/T will be seen to be an important one and is known as the *entropy change.* Most heat transfers, however, occur over a temperature range and the entropy change is defined as

$$\sum_{T_2}^{T_1} \frac{\delta Q}{T} \underset{\text{Reversible}}{} \quad \text{or} \quad \int_{T_1}^{T_2} \frac{dQ}{T} \underset{\text{Reversible}}{}$$

Entropy is given the symbol S, and the entropy change may be written

$$S_2 - S_1 = \int_{T_1}^{T_2} \frac{dQ}{T} \underset{\text{Reversible}}{}$$

where dQ is the infinitesimally small heat transfer at the temperature T degrees absolute.

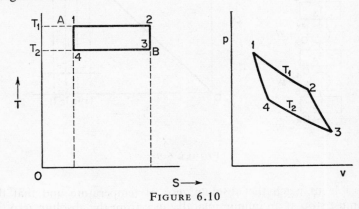

FIGURE 6.10

For a reversible cycle, equation (6.13) may be written in the form

$$\sum \frac{\delta Q}{T} = 0 \qquad (6.14)$$

where $\sum$ means the sum of $\delta Q/T$ around the reversible cycle, or in the form

$$\oint \frac{dQ}{T} \underset{\text{Reversible}}{} = 0 \qquad (6.15)$$

If the entropy change $\delta S = \delta Q/T$ is plotted against absolute temperature T the Carnot cycle can be shown on a temperature–entropy graph as in Fig. 6.10 (a P–V diagram is shown for comparison).

During the isothermal change from 1 to 2, the entropy change $= Q_1/T_1$. During the adiabatic change from 2 to 3 the entropy change $= 0$. During the isothermal change from 3 to 4, the entropy change $= Q_2/T_2$ (this change is negative since Q_2 is heat rejected).

During the adiabatic change from 4 to 1, the entropy change = 0. Therefore

$$\sum \frac{Q}{T} \text{ (from A to B and back to A)} = 0$$

i.e.
$$\sum_{A}^{B} \frac{Q}{T} = \sum_{B}^{A} \frac{Q}{T} \qquad (6.16)$$

or the entropy change from A to B equals entropy change from B to A. This means that entropy change is independent of the path taken and that entropy is, therefore, a property of the system. It must be noted that $\int dQ/T$ for an irreversible process is not the entropy change. However, since entropy is a property, if the end points A and B for an irreversible process are known the entropy change, $S_B - S_A$ can be determined. To do this, convenient reversible processes joining A and B are assumed and $\int dQ/T$ calculated for the assumed processes. Since entropy is a property, this change will be equal to that for irreversible processes between A and B.

6.9 Clausius Inequality

Consider an irreversible engine working side by side with a reversible engine as in Fig. 6.11.

Let η_I = efficiency of irreversible engine

$$= 1 - \frac{\delta Q_2{}'}{\delta Q_1}$$

η_R = efficiency of reversible engine

$$= 1 - \frac{\delta Q_2}{\delta Q_1} = 1 - \frac{T_2}{T_1}$$

Now, if $\eta_I < \eta_R$

then $\delta W' < \delta W$

and $\delta Q_2{}' > \delta Q_2$

Also $1 - \dfrac{\delta Q_2{}'}{\delta Q_1} < 1 - \dfrac{T_2}{T_1}$

and $\dfrac{\delta Q_2{}'}{\delta Q_1} > \dfrac{T_2}{T_1}$

Hence $\dfrac{\delta Q_2{}'}{T_2} > \dfrac{\delta Q_1}{T_1}$

and $\dfrac{\delta Q_1}{T_1} - \dfrac{\delta Q_2{}'}{T_2} < 0$

Writing the sign of $\delta Q_2'$ as negative since $\delta Q_2'$ is negative with respect to δQ_1

$$\frac{\delta Q_1}{T_1} - \frac{(-\delta Q_2')}{T_1} \leqslant 0$$

or

$$\frac{\delta Q_1}{T_1} + \frac{\delta Q_2'}{T_2} \leqslant 0$$

or

$$\sum \frac{\delta Q}{T} \leqslant 0 \text{ for a irreversible cycle} \qquad (6.17)$$

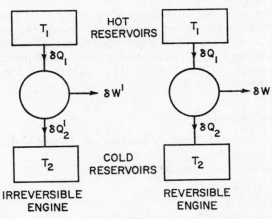

FIGURE 6.11

This equation is known as the Inequality of Clausius and is useful as a criterion of irreversibility in a cycle.

In a reversible cycle $\sum \delta Q/T = 0$. In an irreversible cycle $\sum \delta Q/T < 0$. If, in a proposed cycle $\sum \delta Q/T > 0$, then the cycle would violate the Second Law and would, therefore, be impossible.

6.10 The Temperature–Entropy Diagram

Since entropy is a property of a system, it may be used as a co-ordinate, with temperature as the other ordinate, in order to represent various cycles graphically. It is particularly useful when depicting reversible processes since the areas under the graph represents heat energy transfer. It is of extreme importance to note that, if the process is irreversible, the areas do not represent heat energy transfer. Consider the reversible isothermal and adiabatic processes.

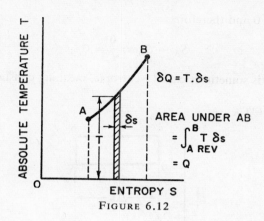

FIGURE 6.12

Isothermal process

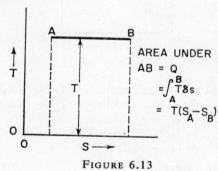

FIGURE 6.13

Reversible adiabatic process (isentropic process)

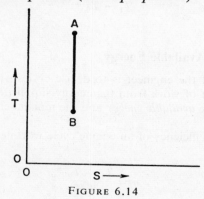

FIGURE 6.14

Here $Q = 0$ and therefore

$$S_A - S_B = \frac{0}{T} = 0$$

This process is sometimes called isentropic, meaning constant entropy.

The Carnot cycle

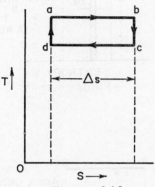

FIGURE 6.15

The heat supplied Q_1 = area under ab = $T_1(S_b - S_a)$
The heat supplied Q_2 = area under cd = $T_2(S_c - S_d)$
Net work transfer = $Q_1 - Q_2$

$$= T_1(S_b - S_a) - T_2(S_c - S_d)$$
$$= (T_1 - T_2)\Delta S$$

Carnot efficiency = $\dfrac{W/J}{Q_1} = \dfrac{(T_1 - T_2)\Delta S}{T_1 \Delta S}$

$$= \frac{T_1 - T_2}{T_1} \qquad (6.18)$$

6.11 Maximum Available Energy

The problem of the engineer is to devise cycles which will give the maximum amount of work from the energy supplied. This maximum work is termed the *available energy* and the remainder, the unavailable energy.

The maximum efficiency of an engine cycle working between T_1 and T_2 is given by

$$\eta = \frac{T_1 - T_2}{T_1}$$

It is obvious that, if this efficiency is to have a high value, T_1 must be made as large as possible and T_2 as small as possible. T_1 is limited by the maximum temperature which metals can withstand and T_2 by the temperature of the surrounding atmosphere. Now since

$$\frac{\delta W_{max}}{J\delta Q_s} = \frac{T_1 - T_2}{T_1}$$

$$\frac{\delta W_{max}}{J} = \delta Q_s \frac{T_1 - T_2}{T_1}$$

$$= \delta Q_s \left(1 - \frac{T_2}{T_1}\right)$$

$$= \delta Q_s - \frac{\delta Q_s}{T_1} T_2$$

$$= \delta Q_s - \delta S T_2 \qquad (6.19)$$

This is represented on a T–S diagram in Fig. 6.16.

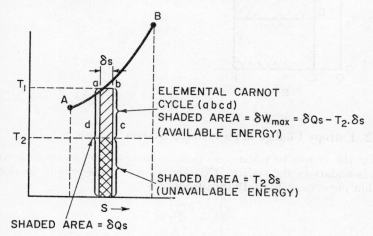

FIGURE 6.16

If the engine received heat during AB which is made up of a number of Carnot cycles then

$$\frac{W_{max}}{J} = \int_A^B \frac{dW_{max}}{J} = \int_A^B dQ_s - \int_A^B T_2 \delta S$$

$$= Q_s - T_2(S_B - S_A) \qquad (6.20)$$

In diagrammatic form this equation may be represented as in Fig. 6.17. Available energy = area ABCD.

The availability for conversion of heat energy into work can be shown to be increased as the initial temperature T_1 is increased since

$$\frac{\delta W_{max}}{J} = \delta Q_s \left(1 - \frac{T_2}{T_1}\right)$$

For a given value of T_2, the ratio of T_2/T_1 will decrease as T_1 is increased. Therefore, δW_{max} will increase as T_1 is increased. From this it is seen that the value of heat is increased as T_1 is increased.

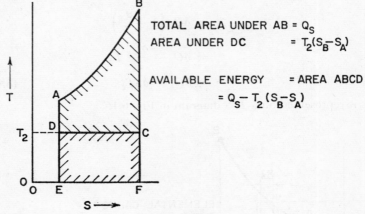

TOTAL AREA UNDER AB = Q_S

AREA UNDER DC = $T_2(S_B - S_A)$

AVAILABLE ENERGY = AREA ABCD

= $Q_S - T_2(S_B - S_A)$

FIGURE 6.17

6.12 Entropy Change in an Isolated System

Let the system be isolated so that no energy transfers occur across the boundary of the system. Consider an irreversible engine working within the system, Fig 6.18 then from §6.9

$$\frac{\delta Q_1}{T_1} - \frac{\delta Q_2'}{T_2} \leqslant 0$$

or

$$\frac{\delta Q_2'}{T_2} > \frac{\delta Q_1}{T_1}$$

Therefore, the entropy gained by the low temperature receiver is greater than that lost by the hot source. The gain in entropy of the isolated system is, therefore

$$\frac{\delta Q_2'}{T_2} - \frac{\delta Q_1}{T_1}$$

This means that all irreversible changes in a system tend to states of greater entropy. This is the Principle of Increase of Entropy.

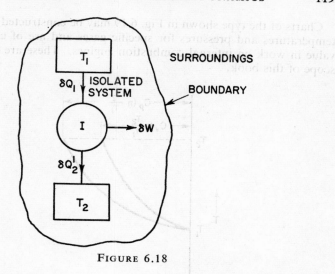

FIGURE 6.18

6.13 Change of Entropy

The heat supplied to a unit mass of fluid for a temperature rise of $\delta T = \delta Q = c\delta T$, where c = specific heat and the specific entropy change

$$s = \frac{\delta Q}{T} = \frac{c\delta T}{T}$$

The total change of specific entropy from T_1 to $T_2 = \int_{T_1}^{T_2} \frac{c\,\mathrm{d}T}{T}$

or

$$s_2 - s_1 = c \ln \frac{T_2}{T_1}$$

If the change takes place at constant pressure

$$s_2 - s_1 = c_\mathrm{p} \ln \frac{T_2}{T_1}$$

If the change takes place at constant volume

$$s_2 - s_1 = c_\mathrm{v} \ln \frac{T_2}{T_1}$$

If these changes are plotted on a temperature–entropy graph the result will be as shown in Fig. 6.19. The entropy change at constant pressure will be larger than at constant volume between given temperatures since $c_\mathrm{p} > c_\mathrm{v}$.

5

Charts of the type shown in Fig. 6.19 may be constructed for various temperatures and pressures for specific gases and are of considerable value in work on internal combustion engines. These are beyond the scope of this book.

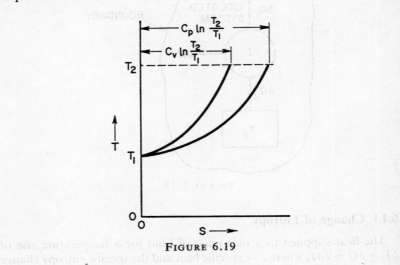

FIGURE 6.19

6.14 Entropy Chart for Steam

For steam tables the zero of entropy is arbitrarily taken as being that of saturated water at $32°F$ or $492°R$, as is the case for all other properties of steam.

Let T_s = absolute saturation temperature
T_{su} = absolute temperature of superheated steam

For unit mass of fluid, the heat added to the liquid

$$\delta Q = c\delta T$$

and the entropy change, $\delta s = \dfrac{c\delta T}{T}$

The entropy change from $492°R$ to T_s

$$s_{T_s} - s_{492} = \int_{492}^{T_s} \frac{c\delta T}{T} = c(\ln T_s - \ln 492) = c \ln \frac{T_s}{492}$$

Now $s_{492} = 0$, s_{T_s} of liquid is given the symbol s_f and $c \simeq 1$

$$\therefore \qquad s_f = \ln \frac{T_s}{492} \qquad (6.21)$$

During the change of phase from liquid to dry saturated vapour the temperature remains constant at T_s and the heat added $= h_g - h_f = h_{fg}$

$$\therefore \qquad s_g - s_f = \frac{h_{fg}}{T_s}$$

where s_g = entropy of dry saturated vapour.
This is sometimes termed evaporation entropy.
  If the steam of dryness fraction x is produced then

$$s_g - s_f = \frac{x h_{fg}}{T_s} \qquad (6.22)$$

The total entropy of dry saturated vapour $= s_g = s_f + \dfrac{h_{fg}}{T}$

$$= \ln \frac{T_s}{492} + \frac{h_{fg}}{T_s} \qquad (6.23)$$

The total entropy of steam of dryness fraction x

$$= \ln \frac{T_s}{492} + \frac{x h_{fg}}{T_s} \qquad (6.24)$$

If the steam is now superheated at constant pressure (as is the case in conventional power generation) then the change in entropy during superheating is

$$s_{gs} - s_g = c_p \ln \frac{T_{su}}{T_s}$$

where s_{gs} is the entropy of superheated steam

$$\therefore \qquad s_{gs} = s_g + c_p \ln \frac{T_{su}}{T_s}$$

$$= \ln \frac{T_s}{492} + \frac{h_{fg}}{T_s} + c_p \ln \frac{T_{su}}{T_s} \qquad (6.25)$$

The value of s_{gs} obtained from equation (6.25) is unreliable since c_p for superheated steam varies considerably with temperature and pressure. Values of s_f, s_g, and s_{gs} are tabulated in steam tables. The temperature–entropy (T–s) chart for steam is shown in Fig. 6.20.

Liquid line
  By plotting values of s_f for various values of T_s the line ABB′P is obtained. Along this line the fluid exists as a saturated liquid.

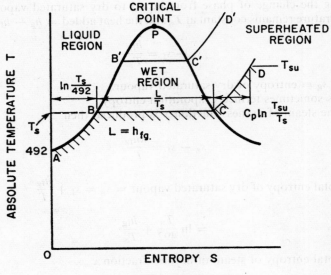

FIGURE 6.20

Vapour line

By plotting lines BC, B'C, etc., corresponding to the evaporation entropy at the various saturation temperatures and then joining points CC', etc., the saturated vapour line PC'C is obtained. At P the critical point, the latent heat is zero and the liquid and vapour coexist.

If points D, D' are obtained for various degrees of superheat $T_{su}-T_s$ the line of constant pressures CD and CD' are obtained for the super-heat field.

The region to the left of the line AP is known as the *liquid field*; that enclosed between the liquid and vapour lines, the *wet field* and that to the right of the vapour line as the *superheated field*.

Lines of constant dryness fraction

If steam of dryness fraction x is produced then the evaporation entropy equals xh_{fg}/T. In Fig. 6.21 the point E represents the state of steam of dryness fraction x and

$$\frac{BE}{BC} = \frac{xh_{fg}/T_s}{h_{fg}/T_s} = x$$

If several points EE', etc., are obtained at various temperatures to represent states of dryness fraction x then the line joining EE', etc., gives a line of constant dryness.

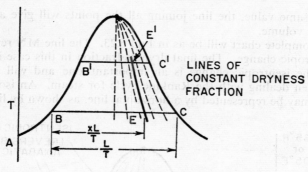

FIGURE 6.21

Lines of constant volume

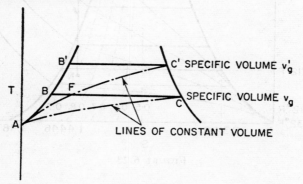

FIGURE 6.22

Consider a point F on the line BC having the same volume as the specific volume of dry vapour at C'. The specific volume of dry vapour at C is equal to v_g. The specific volume at C is greater than that at C' since the pressure there is lower.

Suppose that the dryness fraction at F is x, then the volume at F

$$= v_f + x(v_g - v_f)$$

Since line AFC' is chosen to be a line of constant volume

$$v_f + x(v_g - v_f) = v_g'$$

and

$$x = \frac{v_g' - v_f}{v_g - v_f}$$

If we choose several points corresponding to F at different temperatures so that at each point the dryness fraction, given by the above equation, has the same value, the line joining all the points will give a line of constant volume.

The complete chart will be as in Fig. 6.23. The line MN represents an isentropic change. The final dryness fraction in this case is about 0·88. The isentropic change is an important one and will be used later when dealing with the Rankine cycle for steam. An isothermal change may be represented by a horizontal line, as shown by line OM.

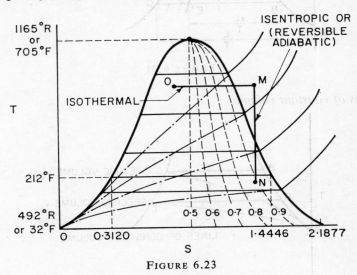

FIGURE 6.23

Ex. 6.4. Determine the entropy of saturated liquid and of saturated steam at 100 lbf/in². Compare the values with those obtained from steam tables.

At 100 lbf/in², $T_s = 327·8 + 460 = 787·8°R$

$$s_f = \ln \frac{T_s}{492} = \ln \frac{787·8}{492} = 0·4713$$

From tables, $s_f = 0·4740$

$$s_g = \ln \frac{T_s}{492} + \frac{h_{fg}}{T_s}$$
$$= 0·4713 + \frac{888·8}{787·8}$$
$$= 0·4713 + 1·1286$$
$$= 1·5999$$

From tables, $s_g = 1·6026$

Ex. 6.5. If the steam in the previous example has a temperature of 750°F, calculate its entropy.

Here
$$s_{gs} = s_g + c_p \ln \frac{T_{su}}{T_s}$$
$$T_{su} = 750 + 460$$
$$= 1,210°R$$

∴
$$s_{gs} = 1·6026 + c_p \ln \frac{1,210}{787·8}$$

c_p is not given. An average value may be obtained from

$$h_{800} - h_{700} = c_p(800 - 700)$$

∴
$$c_p \simeq \frac{h_{800} - h_{700}}{100} = \frac{1,428·9 - 1,378·9}{100}$$

$$= \frac{50}{100} = 0·50$$

$$\therefore \quad s_{gs} = 1·6026 + (0·5 \times 0·4308)$$
$$= 1·8180$$

From tables,

$$\text{at } 800°F \ s_{gs} = 1·8443$$
$$\text{at } 700°F \ s_{gs} = 1·8029$$

for a difference of 100 deg F, $\Delta s_{gs} = 0·0414$
for a difference of 50 deg F, $\Delta s_{gs} = 0·0207$

∴
$$\text{at } 750°F, s_{gs} = 1·8029 + 0·0207$$
$$= 1·8236$$

Ex. 6.6. Steam at 500 lbf/in² and 700°F is expanded isentropically to a pressure of 1 lbf/in². Find the final state of the steam.

From steam tables

$$s_1 = 1·3044$$
$$s_2 = x_g s_2 + (1 - x)s_{f2}$$
$$= x \times 1·9782 + (1 - x)0·1326$$
$$= x1·8456 + 0·1326$$

Now
$$s_2 = s_1, \text{ for an isentropic change}$$

∴
$$x \times 1·8456 + 0·1326 = 1·3044$$

$$x = \frac{1·3044 - 0·1326}{1·8456}$$

$$= \frac{1·1718}{1·8456}$$

$$= 0·635$$

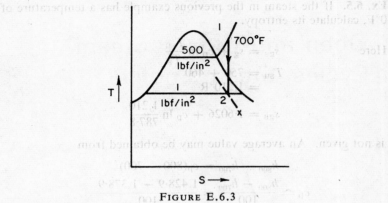

FIGURE E.6.3

EXERCISES ON CHAPTER 6

N.B. All pressures are absolute.

1. An engine working on the Carnot cycle has maximum and minimum temperatures of 2,500°F and 600°F. Determine its efficiency and the heat supply per minute when the output is 30 h.p.

(64·3%; 1,980 Btu)

2. The efficiency of a Carnot cycle is 60%. The minimum temperature in the cycle is 50°F for a perfect gas. Determine:

(a) The maximum temperature in the cycle
(b) The adiabatic expansion ratio for the gas

$$\gamma = 1·4$$

(815°F; 9·9)

3. Discuss the term 'reversibility' and explain why all real processes are never reversible.

4. A refrigerator working on the reversed Carnot cycle has an input work of 5 hp. Determine the heat extracted from the cold chamber in Btu/min if minimum and maximum temperatures in the cycle are 60°F and 100°F.

(2,756 Btu)

5. Determine the specific entropy of steam at 500 lbf/in² if it is (a) 0·85 dry, (b) dry saturated, (c) superheated to 800°F.

(1·3407; 1·4634; 1·6571)

6. Steam at a dryness of 0·85 at a pressure of 40 lbf/in² expands to a pressure of 15 lbf/in² at constant volume. Determine the final specific entropy.

(0·8035)

7. Steam at 100 lbf/in² and 450°F expands isentropically to a pressure of 15 lbf/in². Determine the final state of the steam.

(0·948)

7: Open and Closed Steam Cycles. Steam Engine

7.1 Simple Closed Cycle Steam Power Plant

The components of a simple closed cycle steam plant, illustrated in Fig. 7.1, are:

(a) The boiler, where the water is converted into steam at a constant pressure and temperature by the heat energy received from the combustion of the fuel

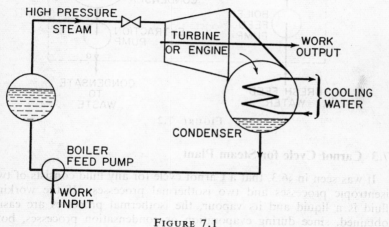

FIGURE 7.1

(b) The engine or turbine, in which the steam expands to a low pressure causing work energy to be available
(c) The condenser, in which heat energy flows from the low pressure steam into the condenser cooling water, resulting in the steam being condensed
(d) The feed pump, which returns the water into the boiler

7.2 Open Steam Cycle

The components of an open cycle are exactly similar to those of the closed; the only difference is that the condensate is discharged to

127

waste and fresh water is pumped into the boiler. This cycle is illustrated
in Fig. 7.2. Such a plant is extremely wasteful on many counts since
both the condensate and the energy contained in it are discharged to a
sink. The water supply to the boiler requires treatment if deposits of
hard scale in the boiler are to be avoided. Such arrangements are
often used in tests on small plant in order to weigh the condensate
at d.

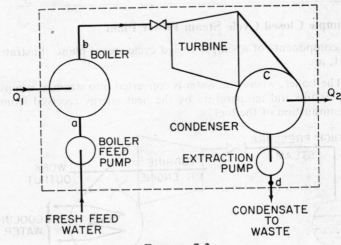

FIGURE 7.2

7.3 Carnot Cycle for Steam Plant

It was seen in §6.3, that a Carnot cycle for any fluid consists of two
isentropic processes and two isothermal processes. If the working
fluid is a liquid and its vapour, the isothermal processes are easily
obtained, since during evaporation or condensation processes, both
the pressure and temperature remain constant. The diagrams shown in
Fig. 7.3 represent a steam plant operating on a Carnot cycle.

Stage 1. a–b
The heat energy is supplied to the boiler resulting in evaporation of
the water, the temperature remaining constant.

Stage 2. b–c
Isentropic expansion takes place in the turbine or engine from b to c.

Stage 3. c–d
In the condenser, condensation takes place from c to d, the temperature remaining constant.

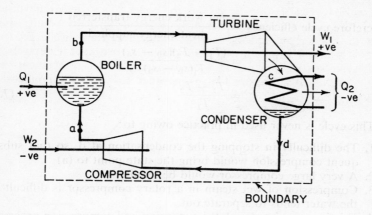

ORGANS OF THE PLANT

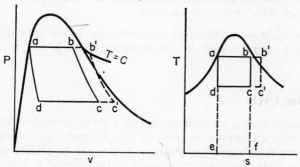

CARNOT CYCLE ON THE Pv AND Ts DIAGRAMS

FIGURE 7.3

Stage 4. d–a

Isentropic compression of the wet steam in a compressor to return the steam to its initial state. The processes are shown on the P–v and T–s diagrams.

From the T–s diagram:

$$\text{heat energy supplied} = Q_1 = \text{area abfe} = T_1(s_b - s_a)$$
$$\text{heat energy rejected} = Q_2 = \text{area cdef} = T_2(s_b - s_a)$$

Therefore net work energy transferred $= \dfrac{W_1}{J} - \dfrac{W_2}{J}$

$$= Q_1 - Q_2$$
$$= (T_1 - T_2)(s_b - s_a)$$

Therefore cycle efficiency $= \dfrac{\text{net work energy transferred}}{\text{heat energy supplied}}$

$$= \frac{(T_1 - T_2)(s_b - s_a)}{T_1(s_b - s_a)}$$

$$= \frac{T_1 - T_2}{T_1} \qquad (7.1)$$

This cycle is never used in practice owing to:

1. The difficulty in stopping the condensation at d, so that subsequent compression would bring the state point to (a)
2. A very large compressor would be required
3. Compression of wet steam in a rotary compressor is difficult as the water tends to separate out
4. Friction associated with the expansion and compression processes would cause the net work done to be very small compared with the work done in the turbine itself

The Carnot cycle is modified to overcome the above difficulties and this modified cycle, known as the Rankine cycle, is widely used in practice.

7.4 Rankine Cycle

A flow diagram of a plant operating on a Rankine cycle is shown in Fig. 7.4. From a comparison between Fig. 7.3 and Fig. 7.4, the similarity between the Carnot and Rankine cycles can be clearly seen. In the Rankine cycle, isentropic expansion in the turbine is followed by the exhaust steam being completely condensed into water in the condenser. This water is then pumped into the boiler by a boiler feed pump. After the feed pump, since the water is not at the saturation temperature corresponding to the pressure, some of the heat energy supplied in the boiler is taken up by the water as sensible heat before evaporation can begin. This results in the boiler process being no longer completely isothermal; the process is, therefore, irreversible, causing the Rankine cycle to be an irreversible cycle and to have a lower efficiency than the Carnot cycle.

At state 1, the high pressure, high temperature steam with a specific enthalpy h_1 leaves the boiler and passes to the engine or turbine. After passing through the turbine and giving up some energy (which is converted to work) the steam leaves the engine or turbine at a low pressure and specific enthalpy h_2 at 2. In the condenser the steam gives up more energy which is removed by the condenser cooling water and in so doing is condensed into water, usually referred to as condensate. The specific enthalpy of the condensate leaving the condenser at state 3 is h_3, and after being raised to a higher pressure at state 4 by the feed pump, the specific enthalpy of the feed is h_4. Consider a steam flow rate of m lb/min circulating round the cycle.

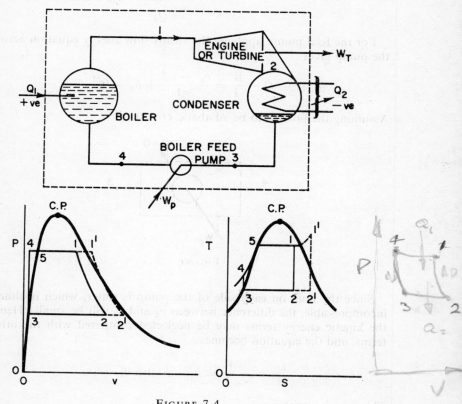

FIGURE 7.4

For the boiler,

heat energy supplied per minute $= (h_1 - h_4)(\text{Btu/lb}) \times m(\text{lb/min})$
$= m(h_1 - h_4)(\text{Btu/min})$
$= Q_1$

For the turbine,

work energy produced per minute $= (h_1 - h_2)(\text{Btu/lb}) \times m(\text{lb/min})$
$= m(h_1 - h_2)(\text{Btu/min})$
$= W_T$

For the condenser,

heat energy removed per minute $= (h_2 - h_3)(\text{Btu/lb}) \times m(\text{lb/min})$
$$= m(h_2 - h_3)(\text{Btu/min})$$
$$= Q_2$$

For the feed pump, applying the steady flow energy equation across the pump gives

$$Q - \frac{W}{J} = \frac{c_4{}^2 - c_3{}^2}{2gJ} + h_4 - h_3$$

Assuming the process to be adiabatic $Q = 0$.

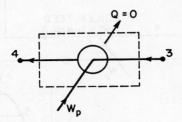

FIGURE 7.5

Since the fluid on each side of the pump is water, which is almost incompressible, the difference between c_4 and c_3 will be small. Hence the kinetic energy terms may be neglected compared with the other terms, and the equation becomes

$$-\frac{W}{J} = h_4 - h_3 \text{ Btu/lb}$$

This may be written as

$$\left(u_4 + \frac{P_4 v_4}{J}\right) - \left(u_3 + \frac{P_3 v_3}{J}\right)$$

$$= (u_4 - u_3) + \frac{P_4 v_4 - P_3 v_3}{J} \text{ Btu/lb}$$

It was stated in chapter 4 that the properties of a liquid do not vary greatly for changes in pressure, i.e. $u_4 \backsimeq u_3$, $v_4 \backsimeq v_3$

Hence the equation becomes

$$-\frac{W}{J} = \frac{(P_4 - P_3)v_3}{J} \text{ Btu/lb} \qquad (7.2)$$

Therefore for a flow rate of m lb/minute, feed pump work energy required per minute

$$= \frac{W_p}{J} = (h_4 - h_3)(\text{Btu/lb}) \times m(\text{lb/min})$$

$$= m(h_4 - h_3) \text{ Btu/min}$$

$$\simeq m \frac{(P_4 - P_3)v_3}{J} \text{ Btu/min} \qquad (7.3)$$

Summarising these expressions:

$$Q_1 = m(h_1 - h_4) \text{ Btu/min}$$

$$\frac{W_T}{J} = m(h_1 - h_2) \text{ Btu/min}$$

$$Q_2 = m(h_2 - h_3) \text{ Btu/min}$$

$$\frac{W_p}{J} = m(h_4 - h_3) \text{ Btu/min}$$

Equating the total energy input to the system to the total energy output from the system gives

$$Q_1 + \frac{W_p}{J} = Q_2 + \frac{W_T}{J}$$

i.e. $$Q_1 - Q_2 = \frac{W_T}{J} - \frac{W_p}{J}$$

or algebraic sum of Q = algebraic sum of W

i.e. $$\sum Q - \sum \frac{W}{J} = 0$$

This is an expression of the First Law of Thermodynamics.

The efficiency of the cycle which is known as the Rankine efficiency η is given by

$$\eta = \frac{(W_T/J) - (W_p/J)}{Q_1}$$

$$= \frac{m(h_1 - h_2) - m(h_4 - h_3)}{m(h_1 - h_4)}$$

$$= \frac{(h_1 - h_2) - (h_4 - h_3)}{(h_1 - h_4)} = \frac{(h_1 - h_2) - (h_4 - h_3)}{(h_1 - h_3) - (h_4 - h_3)} \qquad (7.4)$$

It was shown in (7.3) that

$$(h_4 - h_3) = \frac{(P_4 - P_3)v_3}{J} \qquad (7.5)$$

Compared with $h_1 - h_2$ and $h_1 - h_3$ this quantity is very small and may be neglected. Hence the Rankine efficiency is

$$\eta = \frac{h_1 - h_2}{h_1 - h_3} \tag{7.6}$$

Before this equation can be used, the enthalpy after isentropic expansion h_2 must be determined. This may be done by equating the entropy at state 1 to the entropy at state 2.

i.e. $\qquad\qquad s_1 = s_2$

but $\qquad\qquad s_2 = (1 - x)s_{f_2} + xs_{g_2}$

$\therefore \qquad\qquad s_1 = (1 - x)s_{f_2} + xs_{g_2}$

Knowing s_1, s_{f_2} and s_{g_2}, one can determine x from this equation and h_2 from the relationship

$$h_2 = (1 - x)h_{f_2} + xh_{g_2}$$

Ex. 7.1. In the plant illustrated in Fig. 7.4 the steam leaves the boiler at 300 lbf/in² with 120 degF of superheat. The exhaust pressure is 1 lbf/in² and the condition of the steam entering the condenser is found to be 0·95 dry. The condensate leaves the condenser at the saturation temperature corresponding to 1 lbf/in².

Determine Q_1, Q_2, W_T and W_p in Btu/lb.

At 300 lbf/in² and 120 degF of superheat

$$h_1 = 1{,}278{\cdot}9 \text{ Btu/lb}$$

At 1 lbf/in² and 0·95 dryness

$$\begin{aligned}
h_2 &= (1 - x)h_f + xh_g \\
&= 0{\cdot}05 \times 69{\cdot}7 + 0{\cdot}95 \times 1{,}106 \\
&= 3{\cdot}5 + 1{,}052 \\
&= 1{,}055{\cdot}5 \text{ Btu/lb}
\end{aligned}$$

Enthalpy of condensate, $h_3 = 69{\cdot}7$ Btu/lb

$$\begin{aligned}
\text{Feed pump work} &= \frac{(P_4 - P_3)v_f}{J} \\
&= \frac{144(300 - 1) \times 0{\cdot}016}{778} \\
&= 0{\cdot}89 \text{ Btu/lb}
\end{aligned}$$

Enthalpy of feed, $h_4 = h_3 + \dfrac{(P_4 - P_3)v_f}{J}$

$$= 69\cdot7 + 0\cdot89$$
$$= 70\cdot59 \text{ Btu/lb}$$

$Q_1 = h_1 - h_4 = 1{,}279\cdot6 - 70\cdot59 = 1{,}208\cdot31 \text{ Btu/lb}$

$Q_2 = h_2 - h_3 = 1{,}055\cdot5 - 69\cdot7 = 985\cdot8 \text{ Btu/lb}$

$\Sigma Q = Q_1 + (-Q_2)$ 　　　　　　　$222\cdot51 \text{ Btu/lb}$

$\dfrac{W_T}{J} = h_1 - h_2 = 1{,}278\cdot9 - 1{,}055\cdot5 = 223\cdot4 \;\; \text{Btu/lb}$

$\dfrac{W_p}{J} = h_4 - h_3$ 　　　　　　　　$= 0\cdot89 \text{ Btu/lb}$

$\Sigma \dfrac{W}{J} = \dfrac{W_T}{J} + \left(-\dfrac{W_p}{J}\right)$ 　　　$= 222\cdot51 \text{ Btu/lb}$

Hence 　　　　　　$\Sigma Q - \dfrac{\Sigma W}{J} = 0$

Ex. 7.2. A steam turbine plant is supplied with steam at 600 lbf/in² and 500°F, and exhausts into a condenser at 1 lbf/in². Determine:

(a) The dryness fraction of the steam entering the condenser assuming isentropic expansion

(b) The Rankine efficiency, neglecting the feed pump work

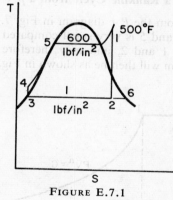

FIGURE E.7.1

From tables

$h_1 = 1{,}215\cdot7 \text{ Btu/lb}$
$s_1 = 1\cdot4586$
$s_2 = s_{f2} + x_2(s_{g2} - s_{f2})$
$ = s_{f2} + x_2 s_{fg}$
$ = 0\cdot1326 + x_2 \times 1\cdot8456$

Now
$$s_2 = s_1$$
$$0.1326 + x_2 \times 1.8456 = 1.4586$$
$$x_2 = \frac{1.4586 - 0.1326}{1.8456}$$
$$= \frac{1.3260}{1.8456}$$
$$= 0.719$$
$$h_2 = h_{f2} + x_2 h_{fg2}$$
$$= 69.7 + 0.719 \times 1,036.3$$
$$= 69.7 + 745$$
$$= 814.7 \text{ Btu/lb}$$

isentropic work done
$$= h_1 - h_2$$
$$= 1,215.7 - 814.7$$
$$= 401.0 \text{ Btu/lb}$$

heat supplied
$$= h_1 - h_3$$
$$= 1,215.7 - 69.7$$
$$= 1,146.0 \text{ Btu/lb}$$

Rankine efficiency
$$= \frac{401.0}{1,146}$$
$$= 0.35 \text{ or } 35\%$$

7.5 Work done on a Rankine Cycle from a p–v Diagram

As can be seen from the P–v diagram in Fig. 7.4 the volume of the water at states 3, 4 and 5 is very small compared with the volume of the steam at states 1 and 2, and may therefore be neglected. The resulting P–v diagram will then be as shown in Fig. 7.6

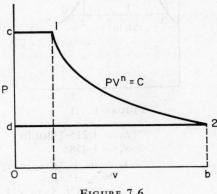

FIGURE 7.6

Let n = isentropic index for steam

then work done = diagram area = $c1a0 + 12ba - d2b0$

$$= P_1 V_1 + \frac{P_1 V_1 - P_2 V_2}{n - 1} - P_2 V_2$$

$$= \frac{P_1 V_1 (n - 1) + (P_1 V_1 - P_2 V_2) - P_2 V_2 (n - 1)}{n - 1}$$

$$= \frac{(n - 1)(P_1 V_1 - P_2 V_2) + (P_1 V_1 - P_2 V_2)}{(n - 1)}$$

$$= \frac{n}{n - 1} (P_1 V_1 - P_2 V_2)$$

$$= \frac{n}{n - 1} P_1 V_1 \left(1 - \frac{P_2 V_2}{P_1 V_1}\right)$$

But $P_1 V_1{}^n = P_2 V_2{}^n$

$$\therefore \qquad \frac{V_2}{V_1} = \left(\frac{P_1}{P_2}\right)^{1/n}$$

$$\therefore \qquad \text{work done} = \frac{n}{n - 1} P_1 V_1 \left\{1 - \frac{P_2}{P_1}\left(\frac{P_1}{P_2}\right)^{1/n}\right\}$$

$$= \frac{n}{n - 1} P_1 V_1 \left\{1 - \left(\frac{P_2}{P_1}\right)^{(n-1)/n}\right\} \qquad (7.7)$$

The objection to the use of this expression lies in the fact that the value of n must be known. The determination of this value is a little laborious. The following values are usually used:

for steam initially superheated, $n = 1\cdot3$
for steam initially saturated, $n = 1\cdot135$

for steam initially having a dryness fraction x, $n = 1\cdot035 + x/10$

7.6 Efficiency Ratio or Relative Efficiency

The actual work done in a steam engine or turbine is less than that given by the reversible adiabatic (isentropic) expansion.

$$\text{The actual thermal efficiency} = \frac{\text{actual work done per lb}}{\text{heat supplied per lb}}$$

$$= \frac{W_a}{h_1 - h_3}$$

$$\text{The Rankine efficiency} = \frac{h_1 - h_2}{h_1 - h_3}$$

$$= \frac{\text{isentropic work done per lb}}{\text{heat supplied per lb}}$$

The efficiency ratio or relative $= \dfrac{\text{actual thermal efficiency}}{\text{Rankine efficiency}}$

$= \dfrac{\text{actual work done per lb}}{\text{heat supplied per lb}} \times \dfrac{\text{heat supplied per lb}}{\text{isentropic work done per lb}}$

$= \dfrac{\text{actual work done per lb}}{\text{isentropic work done per lb}}$ ⠀⠀⠀(7.8)

Ex. 7.3. If, in the previous example, the efficiency ratio of the plant is 0·7 find, (*a*) the thermal efficiency and, (*b*) the steam consumption in lb/kWh.

(*a*) The thermal efficiency $= 0.7 \times 35 = 24.5\%$

(*b*) The actual work done per lb $= 0.7 \times 401 = 280.7$ Btu/lb

$$1 \text{ kWh} = \frac{1,000}{746} \times \frac{33,000 \times 60}{778} = 3,416 \text{ Btu}$$

$$\therefore \text{ Steam consumption} = \frac{3,416}{280.7} = 12.15 \text{ lb/kWh}$$

7.7 Modified Rankine Cycle

If a reciprocating steam engine is used in place of a turbine, the Rankine cycle is modified, as shown in Fig. 7.7. In the reciprocating engine cylinder, the steam is not expanded down to the condenser back

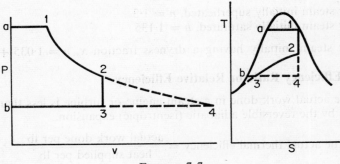

FIGURE 7.7

pressure, but is released before this pressure is reached, with the result that a smaller swept volume is required for a given mass of steam. The amount of work energy lost by this early release (i.e. area 2, 3, 4) is small and the efficiency of the cycle is decreased.

7.8 Reciprocating Steam Engine

The steam engine made possible the industrial revolution and instigated the study of thermodynamics. In transport, in the form of the steam locomotive, it has only recently been superseded, and its place in history, therefore, is one of first importance. In the field of power production, the steam engine now plays a minor part, although it is still employed in small power plants. Fig. 7.8 shows a section

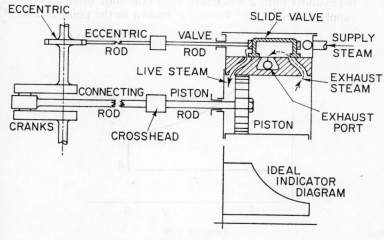

FIGURE 7.8

through the cylinder and valve gear of a double-acting steam engine, the term double acting meaning that steam is admitted to either side of the piston in turn, resulting in two working strokes for each revolution of the crank. Whilst work is being done on one side of the piston, the other side is exhausting. With reference to Fig. 7.6, when the piston is at end A, the slide valve (which is driven by an eccentric mounted on the crankshaft) is arranged to be in such a position that live steam is admitted to end A, and end B is open to exhaust. The live steam forces the piston from left to right causing the crankshaft to rotate. As the piston moves along the cylinder, the slide valves moves from right to left thus cutting off the supply of live steam to end A. The steam which is now contained in the cylinder continues to expand until the piston reaches end B, when the slide valve moves further to the left, resulting in end A being connected to the exhaust, and end B being connected to the live steam. This live steam forces the piston from right to left, and a sequence of events similar to that described above occurs. When the piston reaches A again, the whole cycle has been completed in one revolution of the crankshaft.

7.9 Hypothetical Diagram for a Steam Engine

It is usual to show the ideal indicator diagram for one side of the piston only. The cylinder clearance volume may be neglected. The sequence of events as described above will appear on the indicator diagram as shown (Fig. 7.9).

    1–2. Live steam is admitted to the cylinder forcing the piston down the cylinder until 2 is reached, where the slide valve cuts off the supply of live steam. Point 2 is known as the point of cut-off.

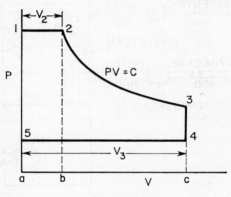

FIGURE 7.9

    2–3. The steam contained in the cylinder continues to expand until the piston reaches the end of its stroke at 3. This process is termed expansive working, and it is usual to assume that this expansion follows a hyperbolic law, $PV = C$; an assumption which is almost true in practice.

    3–4. At 3, the slide valve opens up the cylinder to exhaust, resulting in the pressure of the steam in the cylinder falling instantaneously (i.e. at constant volume) to the condenser pressure or back pressure.

    4–5. As the piston returns along the cylinder, the steam is exhausted at constant pressure.

In Fig. 7.9

work transfer per cycle = area enclosed by the diagram 1 2 3 4 5

$$= \text{area } 1\,2\,ba + \text{area } 2\,3\,cb - \text{area } 5\,4\,ca$$

$$= P_2(V_2 - 0) + P_2 V_2 \ln \frac{V_3}{V_2} - P_b(V_3 - 0)$$

The ratio V_3/V_2 is termed the expansion ratio and is normally denoted by r.

$$\therefore \qquad \text{work transfer per cycle} = P_2 V_2 (1 + \ln r) - P_b V_3$$

If a constant pressure P_m was to act in the cylinder throughout the complete stroke and produce the same work transfer then

$$P_m(V_3 - 0) = P_2 V_2 (1 + \ln r) - P_b V_3$$

$$\therefore \qquad P_m = \frac{P_2 V_2 (1 + \ln r)}{V_3} - P_b$$

$$= \frac{P_2}{r}(1 + \ln r) - P_b \qquad (7.9)$$

P_m is known as the hypothetical mean effective pressure.

7.10 The Actual Indicator Diagram

The indicator diagram (Fig. 7.10) obtained from an actual engine will differ from the ideal indicator diagram in the following ways:

1. There is always a clearance volume in an actual engine
2. Owing to the pressure drop in the steam line and the throttling effect of the valves, the pressure of the steam at entry to the

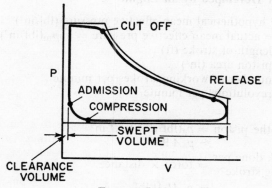

FIGURE 7.10

cylinder is always less than the boiler pressure, and falls slightly until cut-off occurs
3. Cut-off does not appear as a definite point, since valve closure is not instantaneous
4. The actual expansion curve is somewhat lower than the ideal line, owing to condensation of some of the steam on the cylinder walls

5. The steam is released from the cylinder before the end of the stroke, as some time is necessary for the steam pressure to fall to the back pressure. This results in a rounding of the tip of the diagram

6. The back pressure is slightly higher than the condenser pressure during the exhaust stroke

7. Before the end of the exhaust stroke, the exhaust valve is closed, thus trapping a quantity of steam in the cylinder. This steam acts as a cushion for the piston at the end of its stroke, and so reduces the stresses in the piston rod

8. The live steam is supplied to the engine just before the end of the exhaust stroke

The area of the actual diagram is, therefore, smaller than that of the ideal indicator diagram. The ratio of the two areas is taken into account by means of a diagram factor k_d

$$
\begin{aligned}
k_d &= \frac{\text{actual diagram area}}{\text{hypothetical diagram area}} \\
&= \frac{\text{actual mean effective pressure} \times \text{stroke}}{\text{hypothetical mean effective pressure} \times \text{stroke}} \\
&= \frac{\text{actual mean effective pressure}}{\text{hypothetical mean effective pressure}} \qquad (7.10)
\end{aligned}
$$

7.11 Power Developed by an Engine

Let p_m = hypothetical mean effective pressure (lbf/in²)
Then p_a = actual mean effective pressure = $k_d p_m$(lbf/in²)
Let L = length of stroke (ft)
A = piston area (in²)
E = number of working strokes per minute
N = revolutions per minute

Then

$$
\begin{aligned}
\text{force on the piston} &= p_a(\text{lbf/in}^2) \times A \text{ in}^2 \\
&= p_a A \text{ lbf}
\end{aligned}
$$

∴ work done per working stroke = force × distance

$$= p_a AL \text{ ft lbf}$$

∴ work done per minute = $p_a AL$ ft lbf × E working strokes per min

$$= p_a ALE \text{ ft lbf/min}$$

∴ i.h.p. $= \dfrac{p_a ALE}{33,000}$

$$= \frac{k_d p_m ALE}{33,000} \qquad (7.11)$$

For a single acting engine, there will be one working stroke per revolution and $E = N$.

For a double acting engine, there will be two working strokes per revolution and $E = 2N$.

Ex. 7.4. Steam is admitted to a single cylinder double acting engine at 200 lbf/in² and 0·96 dryness fraction. The back pressure is 17 lbf/in². The bore and stroke are 8 in and 10 in respectively. The speed is 250 rev/min.

If cut off takes place at quarter stroke, and the diagram factor is 0·8, determine the iph developed and the weight of steam admitted per minute. Assume hyperbolic expansion of the steam and neglect clearance.

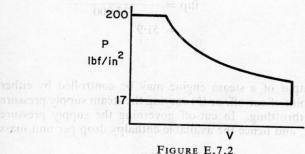

FIGURE E.7.2

$$\text{volume of cylinder} = \frac{\pi}{4} \times \frac{64 \times 10}{1,728}$$

$$= 0\text{·}292 \text{ ft}^3$$

$$\text{volume at cut off} = \frac{0\text{·}292}{4}$$

$$= 0\text{·}073 \text{ ft}^3$$

Specific volume at 200 lbf/in² and 0·96 dryness

$$= 0\text{·}96 \times 2\text{·}29$$

$$= 2\text{·}2 \text{ ft}^3/\text{lb}$$

$$\text{Weight of steam admitted per stroke} = \frac{0\text{·}073}{2\text{·}2}$$

$$= 0\text{·}0332 \text{ lbf}$$

$$\text{weight of steam admitted per min} = 0\text{·}0332 \times 2 \times 250$$

$$= 16\text{·}6 \text{ lbf}$$

$$\text{hypothetical mean effective pressure} = \frac{P_1(1 + \ln r)}{r} - P_b$$

$$= \frac{200(1 + \ln 4)}{4} - 17$$

$$= 102 \cdot 3 \text{ lbf/in}^2$$

$$\text{actual mean effective pressure} = 0 \cdot 8 \times 102 \cdot 3$$

$$= 81 \cdot 84 \text{ lbf/in}^2$$

$$\text{work done per stroke} = 81 \cdot 84 \times \frac{\pi}{4} \times 8^2 \times \frac{10}{12}$$

$$= 3,430 \text{ ft lbf}$$

$$\text{work done per min} = 3,430 \times 2 \times 250$$

$$\therefore \quad \text{ihp} = \frac{3,430 \times 2 \times 250}{33,000}$$

$$= 51 \cdot 9$$

7.12 Governing

The power output of a steam engine may be controlled by either (a) varying the point of cut-off, or (b) varying the steam supply pressure to the chest by throttling. In cut-off governing the supply pressure remains constant, and hence the available enthalpy drop per unit mass

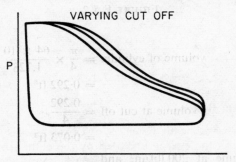

FIGURE 7.11

of steam over the engine is unaffected. The effect of cut-off governing on the indicator diagram is shown in Fig. 7.11. In throttle governing, since the initial pressure of the steam is varied, the available enthalpy drop per unit mass of steam is also varied. The effect of throttle governing of the indicator diagram is shown in Fig. 7.12. Of the two methods of governing, the cut-off method is most economical and is used when the control is carried out manually. Where the engine is speed governed, the throttling method is used. The steam consumption curves for both methods are shown in Fig. 7.13.

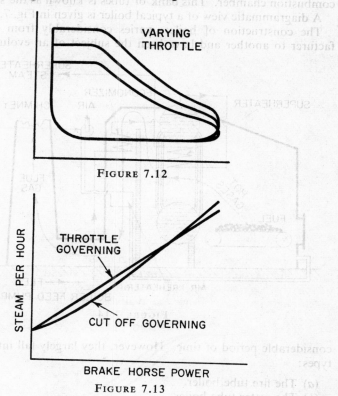

FIGURE 7.12

FIGURE 7.13

7.13 The Steam Boiler

The function of a steam boiler is to convert water into steam at constant pressure, the energy liberated by the combustion of the fuel in the grate being used to increase the enthalpy of the fluid passing through the boiler. Some of the energy liberated tends to leave with the flue gases through the chimney and an attempt is made to regain some of this energy by installing an economiser and an air preheater in the path of the flue gases. In passing through the economiser, the boiler feed water is heated at the expense of the flue gases and therefore enters the boiler with an increased enthalpy. In the air preheater, the air required for the combustion of the fuel is heated, resulting in an increase in temperature of the combustion products.

It is often required that a steam supply be superheated. This is effected by arranging that after the steam leaves the boiler drum, it passes through a bank of tubes situated in the radiant zone of the boiler combustion chamber. This bank of tubes is known as the superheater.

A diagrammatic view of a typical boiler is given in Fig. 7.14.

The construction of boilers varies considerably from one manufacturer to another and have been the subject of an evolution over a

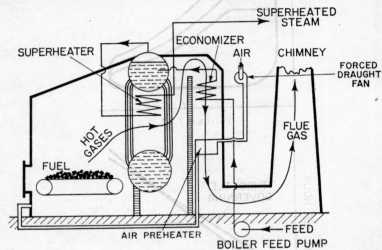

FIGURE 7.14

considerable period of time. However, they largely fall into two main types:

(a) The fire tube boiler
(b) The water tube boiler

In the fire tube boiler, the hot gases pass through tubes surrounded by the water from which the steam is to be generated. Examples of this type of boiler are the Lancashire, Cochran, and Cornish boilers.

In the water tube boiler, the water is made to pass through a large number of small bore tubes over which the hot gases flow. This type of boiler is the most recent in the process of evolution, and most modern boilers are built on this principle.

Resort should be made to books on boilers for detailed description of the various types.

Rating of a boiler

A boiler is rated by the maximum amount of steam generated in pounds per hour, e.g. a boiler supplying a 30,000 kW steam turbine may be rated at 250,000 lb/h.

Equivalent evaporation from, and at, 212°F

Boilers differ widely in their pressures, temperatures, and inlet feed temperatures. In order to compare boilers one against the other, the concept of equivalent evaporation from, and at, 212°F is used.

If a boiler is supplied with feed water at 212°F and generates dry saturated steam at 212°F the energy supplied per lb of steam generated is h_{fg} at 212°F = 970 Btu/lb. The energy required to generate one pound of steam in any boiler is given by $h_s - h_w$ Btu/lb where h_s = enthalpy of steam produced and h_w = enthalpy of feed entering boiler.

If W_a lb/h of steam are produced in such a boiler, then the energy gained by the steam per hour will be $W_a(h_s - h_w)$ Btu. In the 'equivalent' boiler this amount of energy would generate $W_a(h_s - h_w)/970$ lb of steam. This expression is known as the equivalent evaporation from, and at, 212°F.

Boiler efficiency

The efficiency of a boiler is defined as

$$\frac{\text{energy gained by fluid}}{\text{energy input to boiler}}$$

If W_s lb/h = rate at which steam is generated

W_f lb/h = rate at which fuel is supplied

h_s Btu/lb = specific enthalpy of steam generated

h_w Btu/lb = specific enthalpy of feed water

CV Btu/lb = calorific value of the fuel

then

energy gained by fluid = $W_s(h_s - h_w)$ Btu/h

energy supplied to boiler = $W_f(CV)$

and   boiler efficiency = $\dfrac{W_s(h_s - h_w)}{W_f(CV)}$ (7.12)

Ex. 7.5. A steam boiler evaporates 69,440 lb of water per hour at a pressure and temperature of 400 lbf/in² and 685°F respectively. The temperature of the feed to the boiler is 255°F. If the weight of coal fired per hour is 10,100 lb and its calorific value is 11,200 Btu/lb, determine:

(a) The boiler efficiency

(b) The equivalent evaporation from and at 212°F per lb of coal

From tables, the enthalpy of steam at 400 lbf/in² and 685°F = 1,354·4 Btu/lb.

The enthalpy of feed water at $255°F = 223·4$ Btu/lb.

Therefore heat required to generate 1 lb of steam

$$= 1,354·4 - 223·4$$
$$= 1,131·0 \text{ Btu}$$

Heat required per hour $= 69,440 × 1,131·0$ Btu

Energy available in fuel per hour $= 10,100 × 11,200$ Btu

$$\therefore \quad \text{boiler efficiency} = \frac{69,440 × 1,131·0}{10,100 × 11,200} × 100$$

$$= 69·5\%$$

Heat used to generate steam per lb of coal

$$= \frac{69,440 × 1,131}{10,100}$$

$$= 7,780 \text{ Btu}$$

Therefore equivalent evaporation from and at $212°F$

$$= \frac{7,780}{970} = 8·03 \text{ lb}$$

EXERCISES IN CHAPTER 7

N.B. All pressures in the following exercises are absolute.

1. A steam turbine plant is supplied with steam at 700 lbf/in² and 800°F and exhausts at 0·8 lbf/in². Determine:
   (a) The final state of the steam after isentropic expansion
   (b) The isentropic work done
   (c) The Rankine efficiency
Neglect the feed pump work.

(0·795; 514 Btu/lb; 38·4%)

2. If in exercise 1 above, the efficiency ratio of the actual plant is 0·75 determine
   (a) The actual final state of the steam after expansion in the turbine
   (b) The steam consumption in lb/hp h

(0·92; 6·6

3. Explain why the Rankine cycle and not the Carnot cycle is taken as the ideal cycle for steam plant. Sketch the $P–V$ and $T–s$ diagram for both these cycles when using steam as the working fluid.

4. A steam engine works on the modified Rankine cycle. Dry steam is supplied at 120 lbf/in², the expansion ratio is 3, and the exhaust pressure is 15 lbf/in². Determine:
   (a) The dryness fraction at the end of expansion
   (b) The work done per lb of steam supplied
You may assume n in $PV^n = c$ for the isentropic expansion to be 1·135.

(0·923; 135 Btu

5. A single cylinder, double acting steam engine has a bore and stroke of 10 in. It is supplied with dry saturated steam at 200 lbf/in² and exhausts at 10 lbf/in². It runs at 280 rev/min. When the cut-off is at $\frac{1}{4}$ stroke the engine is found to develop 85 b.h.p. Assuming a mechanical efficiency of 80% and neglecting the effect of clearance, determine the diagram factor. Determine, also, the weight of steam supplied per minute. Assume also hyperbolic expansion.

(0·87; 26·9 lbf)

6. A single cylinder, double acting steam engine is to be designed to develop 80 bhp at a speed of 300 rev/min. The boiler pressure is 200 lbf/in² and the exhaust occurs at an atmospheric pressure of 14·7 lbf/in².

Find the cylinder bore and stroke length, if the mechanical efficiency is 85%, the diagram factor is 0·8, cut-off occurs at three-eights of the stroke, and the stroke is 1·5 times the bore.

Assume the expansion to be hyperbolic and neglect clearance lb/bhp/h. Find, also, the theoretical steam consumption in lb/bhp/h if the steam at entry to the cylinder has a dryness fraction of 0·96.

(7·9 in; 11·85 in; 25·8 lb/bhp/h)

7. A single cylinder double acting steam engine is required to develop 130 ihp at 200 rev/min when supplied with steam at 200 lbf/in² and cut-off is to be at 0·25 of the stroke. Back pressure is 20 lbf/in².

Assuming a diagram factor of 0·83, and hyperbolic expansion, calculate the cylinder dimensions if the stroke:bore ratio is 1:4. Clearance and the area of the piston rod may be neglected.

(bore 11·25 in; stroke 15·75 in)

8. A single cylinder double acting steam engine develops 120 bhp at 180 rev/min. Steam is supplied at 180 lbf/in², cut-off occurs at $\frac{1}{3}$ stroke, and the back pressure is 15 lbf/in². If the diagram factor is 0·75 and the mechanical efficiency is 80%, calculate the cylinder dimensions for a stroke:bore ratio of 1·2:1. Hyperbolic expansion may be assumed and clearance and the effect of piston rod area may be neglected.

(12·8 in; 15·35 in)

If the steam in the cylinder is 0·9 dry at the end of admission, calculate the steam consumption of the engine in lb/h.

(3,600 lb/h)

9. A single cylinder double acting steam engine develops 18 bhp at 210 rev/min with a mechanical efficiency of 70%. The engine has a bore of 8 in. and a stroke of 12 in. Steam is supplied at 100 lbf/in² and the back pressure is 10 lbf/in². If the expansion is hyperbolic and clearance and the effect of piston rod area may be neglected, calculate the diagram factor if the cut-off is at 30% of the stroke.

(0·715)

10. Discuss methods of increasing the efficiency of a simple boiler plant. During an 8 hour trial on a boiler the following average data were obtained:

Pressure of the steam leaving the boiler	= 200 lbf/in²
Feed water evaporated	= 58,900 lbf
Temperature of the feedwater at inlet	= 120°F
Weight of coal fired	= 9,400 lbf
Calorific value of coal as fired	= 12,400 Btu/lbf

Air supplied per lbf of coal	=	17 lbf
Temperature of flue gases leaving boiler	=	650°F
Boiler house temperature	=	70°F

Samples of the steam leaving the boiler were passed through a throttling calorimeter and the average value of the pressure and temperature after throttling was 15 lbf/in² and 268°F respectively. Determine:

(a) The condition of the steam leaving the boiler
(b) The boiler efficiency
(c) Equivalent evaporation from, and at, 212°F per lbf of coal
(d) The heat loss in flue gas above 70°F expressed as a percentage of the heat supplied

Assume the specific heat of the flue gases to be 0·25.

(0·975; 55·2%; 7·04; 21·2%)

11. Explain the significance of the term 'from and at 212°F'. A boiler in a power station generates 280,000 lb of steam per hour at a pressure of 500 lbf/in² and 700°F. The feed temperature to the boiler is 400°F. The boiler efficiency is estimated to be 85%. Determine:

(a) The hourly demand of coal of calorific value, 12,500 Btu/lb
(b) The equivalent evaporation from and at 212°F per lb of coal

(25,950 lb; 10·92 lb)

12. A boiler generates steam at 400 lbf/in² and 500°F. It is found that the steam leaving the boiler drum has a dryness fraction of 0·96. The feed water to the boiler enters the economiser at a temperature of 100°F, where it is heated to 180°F. 9·5 lb of steam is generated per lb of coal burned. The coal burned has a calorific value of 13,000 Btu/lb.

Calculate:

(a) The heat supplied in the economiser, boiler drum and superheater per lb of steam
(b) The plant efficiency

(80; 1,026; 71·1 Btu/lb; 86%)

13. Sketch a water-tube boiler circuit showing a superheater and an economiser. Show, clearly, the paths of the water, steam and gases.

An oil-fired boiler, supplied with feed water at 140°F and generating steam at 300 lbf/in² and 500°F is estimated to have an efficiency of 78%. The calorific value of the oil is 18,600 Btu/lb. Estimate the weight of oil required per working day of 16 hours for an evaporation of 4,000 lb of steam per hour.

(5,050 lb)

8: Air Standard Cycles

One of the most common types of heat engine is the internal combustion engine, as used in road transport, rail transport, marine installations, aircraft, stationary plant, etc. The working fluid in the internal combustion engine is mainly air, but the fuels may vary from the light petrols to the heavy fuel oils. Internal combustion engines may be broadly divided into two types as follows:

(a) Those which make use of a series of non-flow processes to convert heat energy into work energy, e.g. reciprocating engines

(b) Those which make use of flow processes to convert heat energy into work energy, e.g. gas turbines

In order to carry out a simplified analysis of the various internal combustion engine cycles, it is assumed that the working fluid is air, that it behaves as a perfect gas, and that there is no change in the composition of the air during the complete cycle. Such cycles are known as Air Standard Cycles and although they utlilise the above assumptions, they do provide close indications as to the behaviour of working fluids in actual internal combustion engines.

8.1 Constant Volume or Otto Cycle

This cycle may be likened to the cycle on which an ordinary petrol engine operates. The cycle itself consists of four non-flow processes which may be described on the air standard cycle as follows:

At 1 the piston is at the outer dead centre (ODC) with the cylinder completely full of air, at a low pressure and temperature.

1–2. Rapid compression of the air (assumed to be adiabatic and reversible, and thus obeying the equation $PV^\gamma =$ constant) by the piston moving quickly into the inner dead centre (IDC). The air is compressed into the clearance volume of the cylinder, thus raising the pressure and temperature of the air.

2–3. Whilst the piston remains at the inner dead centre and the volume of the air remains constant, heat energy is supplied to the air from the surroundings thus further increasing the pressure and the temperature of the air to the maximum values attained in the cycle.

3–4. The hot high pressure air contained in the cylinder at 3 forces the piston rapidly down the cylinder, causing work energy to

be transferred to the surroundings at the expense of the internal
energy of the air. Again, since this is a non-flow process which
takes place rapidly it is assumed to be reversible and adiabatic,
following the equation $PV^\gamma =$ constant. When the piston
reaches ODC at 4 the volume of the air is maintained constant
whilst heat transfer takes place, energy flowing from the air to
the surroundings. This process proceeds (4–1) until the air
reaches its original condition at 1. The cycle is then repeated.

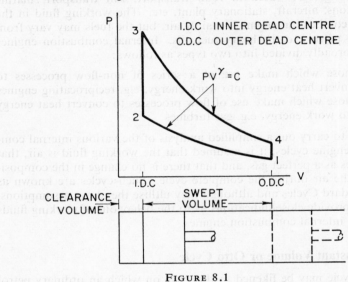

FIGURE 8.1

In an actual engine cylinder, slight departures from the air standard
cycle arise:

(a) Initially, the cylinder contains not air, but a mixture of air, petrol
vapour and residual gases and it is this mixture which is compressed into the clearance volume

(b) The compression process will not be reversible, hence the compression curve will deviate slightly from the compression line
on the air standard cycle

(c) The heat energy supplied to the air during 2–3 is not supplied
from the surroundings but is liberated when a spark causes the
petrol vapour to burn

(d) Although the petrol vapour burns rapidly, it does not do so
instantaneously, with the result that the piston will have started
to move back down the cylinder by the time burning is complete
thus causing the peak of the diagram at 3 to be slightly rounded

(e) On the expansion stroke the cylinder will contain products of combustion. The actual expansion process will not be reversible and this, coupled with changes in the chemical composition of the products (due to the high temperatures involved), causes the expansion line to depart from the theoretical expansion line

(f) The exhaust valve cannot open and close instantaneously, hence the points of the diagram at 4 and 1 become slightly rounded. Also, time is required for the products of combustion to flow out of the cylinder and the new mixture to flow into the cylinder before the commencement of the next cycle

An expression for the efficiency of the air standard Otto cycle may be obtained by considering the cylinder to contain 1 lb of air. Thus, if P, v and T refer to the pressure, specific volume and absolute temperature of the air, and suffixes 1, 2, 3 and 4 refer to the state points as shown in Fig. 8.1 then we have

$$\frac{P_1v_1}{T_1} = \frac{P_2v_2}{T_2} = \frac{P_3v_3}{T_3} = \frac{P_4v_4}{T_4} \tag{8.1}$$

Also, since the process 1 to 2 is a reversible adiabatic

$$P_1v_1{}^\gamma = P_2v_2{}^\gamma$$

$$\therefore \quad \frac{T_2}{T_1} = \frac{P_2v_2}{P_1v_1} = \left(\frac{v_1}{v_2}\right)^\gamma \frac{v_2}{v_1} = \left(\frac{v_1}{v_2}\right)^\gamma \left(\frac{v_1}{v_2}\right)^{-1}$$

$$\text{or} \quad \frac{T_2}{T_1} = \left(\frac{v_1}{v_2}\right)^{\gamma-1} \tag{8.2}$$

Similarly, since the process 3 to 4 is a reversible adiabatic

$$P_3v_3{}^\gamma = P_4v_4{}^\gamma$$

$$\text{and} \quad \frac{T_4}{T_3} = \frac{P_4v_4}{P_3v_3} = \left(\frac{v_3}{v_4}\right)^\gamma \frac{v_4}{v_3} = \left(\frac{v_3}{v_4}\right)^{\gamma-1} \tag{8.3}$$

Now, the ratio $v_1/v_2 = v_4/v_3$ is called the volume compression ratio, r_v

$$\therefore \quad \frac{T_2}{T_1} = r_\mathrm{v}{}^{\gamma-1}$$

$$\text{and} \quad \frac{T_4}{T_3} = \frac{1}{r_\mathrm{v}{}^{\gamma-1}}$$

$$\text{Hence} \quad \frac{T_2}{T_1} = \frac{T_3}{T_4}$$

$$\text{or} \quad \frac{T_4}{T_1} = \frac{T_3}{T_2} \tag{8.4}$$

$$\text{Cycle efficiency} = \frac{\text{net work energy transferred}}{\text{heat energy supplied}}$$

$$= \frac{\text{heat energy supplied} - \text{heat energy rejected}}{\text{heat energy supplied}}$$

$$= 1 - \frac{\text{heat energy rejected}}{\text{heat energy supplied}}$$

Since 1–2 and 3–4 are adiabatic processes, no heat energy transfers will take place during these processes. The amount of heat energy transferred to the air will be given by applying the non-flow energy equation $Q - W = U_2 - U_1$ to the constant volume processes.

Thus, if heat energy supplied during the constant volume process 2–3 is denoted by Q_{23} we have

$$Q_{23} - 0 = U_3 - U_2$$

For 1 lb of perfect gas

$$U_3 - U_2 = 1 \times c_\mathrm{v} \times (T_3 - T_2)$$
$$\therefore \qquad Q_{23} = c_\mathrm{v}(T_3 - T_2) \qquad (8.5)$$

Heat energy is transferred from the air during the process 4–1 and as above we have

$$- Q_{41} - 0 = U_1 - U_4$$
$$= 1 \times c_\mathrm{v}(T_1 - T_4)$$
$$\therefore \qquad Q_{41} = c_\mathrm{v}(T_4 - T_1) \qquad (8.6)$$

$$\therefore \qquad \text{cycle efficiency} = 1 - \frac{\text{heat energy rejected}}{\text{heat energy supplied}}$$

$$= 1 - \frac{Q_{41}}{Q_{23}}$$

$$= 1 - \frac{c_\mathrm{v}(T_4 - T_1)}{c_\mathrm{v}(T_3 - T_2)}$$

$$= 1 - \frac{T_1\{(T_4/T_1) - 1\}}{T_2\{(T_3/T_2) - 1\}}$$

but from equation (8.5)

$$\frac{T_4}{T_1} = \frac{T_3}{T_2}$$

$$\therefore \qquad \frac{T_4}{T_1} - 1 = \frac{T_3}{T_2} - 1$$

$$\therefore \qquad \text{cycle efficiency} = 1 - \frac{T_1}{T_2}$$

$$= 1 - \frac{1}{r_\mathrm{v}^{\gamma-1}} \qquad (8.7)$$

Two important points should be noted with regard to this expression.

(1) The efficiency of a Carnot cycle working between the same temperature limits would be given by $1 - (T_1/T_3)$. Since $T_3 > T_2$, then the efficiency of an Otto cycle must be less than that of a Carnot cycle when working between the same temperature limits.

(2) The above expression shows that the efficiency of an Otto cycle depends only on the compression ratio r_v and γ. Since γ for air has a constant value, then the cycle efficiency will vary only with the compression ratio r_v, which is a mechanical characteristic of the engine, depending only on the clearance and swept volumes of the engine. Thus the efficiency of an engine operating on the air standard Otto cycle will be constant if the compression ratio is constant, i.e. the air standard efficiency (usually abbreviated to A.S.E.) will not vary whatever the load on the engine. Also, all engines having the same compression ratio and operating on this cycle will have the same air standard efficiency.

As discussed earlier, slight departures from the air standard cycle occur in actual engines. In comparing engines of different compression ratios one against the other, relative efficiencies are used, where

$$\text{relative efficiency} = \frac{\text{engine thermal efficiency}}{\text{air standard efficiency}}$$

The relative efficiencies thus give an indication of how the engine efficiency compares with that which is theoretically attainable.

Ex. 8.1. An engine operating on the ideal Otto cycle has a bore of 4 in, a stroke of 5 in, and a compression ratio of $7:1$. At the beginning of the compression stroke the cylinder contains air at $60°F$ and $14·5$ lbf/in². If the maximum cycle temperature is $3,000°F$ determine:

(a) The pressure, volume, and temperature at the cardinal points of the cycle
(b) The air standard efficiency
(c) The mean effective pressure of the cycle

For air take $c_v = 0·1715$ Btu/lb degR and $c_p = 0·24$ Btu/lb degR

$$\text{Swept volume} = V_1 - V_2$$
$$= \left(\frac{\pi}{4} \times 4^2 \times 5\right) \text{ in}^3$$
$$= 62·84 \text{ in}^3 \qquad (a)$$
$$\text{Compression ratio} = \frac{V_1}{V_2} = 7$$
$$\therefore \qquad V_1 = 7V_2$$

Substituting in (*a*) gives

$$V_1 = V_4$$
$$= 73 \cdot 31 \text{ in}^3$$

and

$$V_2 = V_3$$
$$= 10 \cdot 47 \text{ in}^3$$

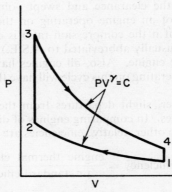

FIGURE E.8.1

Since 1–2 is isentropic, $P_1 V_1{}^\gamma = P_2 V_2{}^\gamma$

where

$$\gamma = \frac{c_p}{c_v} = \frac{0 \cdot 24}{0 \cdot 1715} = 1 \cdot 4$$

$$\therefore \qquad P_2 = \left(\frac{V_1}{V_2}\right)^\gamma P_1$$
$$= 7^{1 \cdot 4} \times 14 \cdot 5 \text{ lbf/in}^2$$
$$= 15 \cdot 3 \times 14 \cdot 5 \text{ lbf/in}^2$$
$$= 222 \text{ lbf/in}^2$$

Also

$$\frac{P_1 V_1}{T_1} = \frac{P_2 V_2}{T_2}$$

$$\therefore \qquad T_2 = \frac{P_2}{P_1} \times \frac{V_2}{V_1} \times T_1$$
$$= \frac{222 \text{ lbf/in}^2}{14 \cdot 5 \text{ lbf/in}^2} \times \frac{1}{7} \times (460 + 60) \,°\text{R}$$
$$= 1{,}137°\text{R}$$

Since
$$\frac{P_2 V_2}{T_2} = \frac{P_3 V_3}{T_3}$$

$\therefore$
$$P_3 = \frac{T_3}{T_2} \times \frac{V_2}{V_3} \times P_2$$
$$= \frac{(3,000 + 460)\,°\text{R}}{1137\,°\text{R}} \times 1 \times 222\,\text{lbf/in}^2$$
$$= 676\,\text{lbf/in}^2$$

Since 3–4 is isentropic
$$P_3 V_3{}^\gamma = P_4 V_4{}^\gamma$$

$\therefore$
$$P_4 = \left(\frac{V_3}{V_4}\right)^\gamma P_3$$
$$= \left(\frac{1}{7}\right)^{1\cdot 4} \times 676\,\text{lbf/in}^2$$
$$= \frac{1}{15\cdot 3} \times 676\,\text{lbf/in}^2$$
$$= 44\cdot 2\,\text{lbf/in}^2$$

Also
$$\frac{P_3 V_3}{T_3} = \frac{P_4 V_4}{T_4}$$

$\therefore$
$$T_4 = \frac{P_4}{P_3} \times \frac{V_4}{V_3} \times T_3$$
$$= \frac{44\cdot 2\,\text{lbf/in}^2}{676\,\text{lbf/in}^2} \times 7 \times 3,460\,°\text{R}$$
$$\doteq 1,580°\text{R}$$

Air standard efficiency $= 1 - \left(\frac{1}{r}\right)^{\gamma - 1}$
$$= 1 - \left(\frac{1}{7}\right)^{1\cdot 4 - 1}$$
$$= 1 - 0\cdot 46$$
$$= 0\cdot 54$$

If m is the mass of air contained in the cylinder. Then at 1
$$m = \frac{P_1 V_1}{R T_1}$$

where $R = J \times (c_p - c_v) = 778 \frac{\text{ft lbf}}{\text{Btu}} \times (0.24 - 0.1715) \frac{\text{Btu}}{\text{lb degR}}$

$= 53.3 \text{ ft lbf/lb degR}$

$\therefore \qquad m = \frac{(14.5 \times 144)(\text{lbf/ft}^2) \times (73.31/1,728) \text{ ft}^3}{53.3 \text{ (ft lbf/lb degR)} \times 520 \text{ degR}}$

$= 0.0032 \text{ lb}$

heat energy supplied per cycle $= mc_v(T_3 - T_2)$

$= 0.0032 \text{ lb} \times 0.1715 \frac{\text{Btu}}{\text{lb degR}} \times (3,460 - 1,137) \text{ degR}$

$= 1.276 \text{ Btu}$

Since cycle efficiency $\eta = \dfrac{\text{net work energy transferred}}{\text{heat energy supplied}}$

$\therefore$ net work energy transferred $= \eta \times$ heat energy supplied

$= 0.54 \times 1.276 \text{ Btu}$

$= 0.54 \times 1.276 \text{ Btu} \times 778 \frac{\text{ft lbf}}{\text{Btu}}$

$= 536 \text{ ft lbf}$

$\therefore$ mean effective pressure $= \dfrac{\text{net work energy transferred}}{\text{volume change}}$

$= \dfrac{536 \text{ ft lbf}}{(62.84/1,728) \text{ ft}^3}$

$= 14,720 \text{ lbf/ft}^2$

$= \dfrac{14,720}{144}$

$= 102.2 \text{ lbf/in}^2$

8.2 Four Stroke and Two Stroke Cycles

It will be seen that the above cycle of events will require two strokes of the piston or one revolution of crankshaft for completion. In a four stroke cycle, Fig. 8.2, i.e. a cycle which requires four strokes of the piston (equal to two revolutions of the crankshaft) the exhaust valve remains open whilst the piston moves back to the IDC to expel the products of combustion. When these have been expelled the exhaust valve closes and the inlet valve opens so that as the piston returns to ODC, thus completing its fourth stroke, a fresh charge of mixture is drawn into the cylinder in order that the cycle may be repeated. It

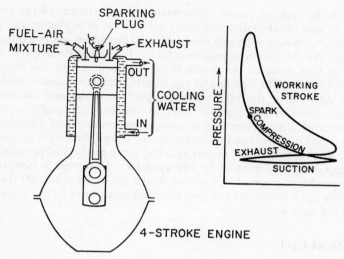

FIGURE 8.2

will be realised that as all these events take place rapidly it is impossible to replace completely the products of combustion with new mixture. Any products remaining in the cylinder at the commencement of the cycle are called residual gases. In a two stroke cycle, Fig. 8.3, i.e. one

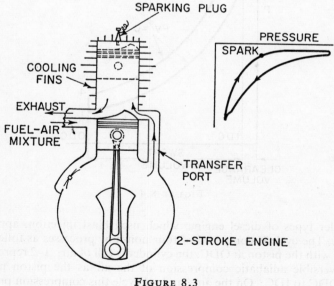

FIGURE 8.3

in which the cycle of events is completed in two strokes of the piston
(equal to one crankshaft revolution) the process of replacing the products of combustion with new mixture has to be carried out in a different
manner. As the piston approaches ODC, the exhaust valve (or port)
opens so that some of the products of combustion flow out into the
exhaust system. The opening of the exhaust valve is quickly followed
by the opening of the inlet valve (or port) and the new mixture is
admitted through the transfer port to the cylinder under a slight
pressure and directed so that it sweeps the products of combustion
before it and out through the exhaust valve before the valves are closed
again. Thus a two-stroke engine requires some method of pressurising
the supply of new mixture to the cylinder. However, a two-stroke
engine produces a power or working stroke for each crankshaft revolution, whereas a four-stroke engine produces one power stroke for every
two crankshaft revolutions.

8.3 Diesel Cycle

It should be noted that this cycle, which is named after its inventor,
is not the cycle on which the modern solid injection diesel engine
operates (see §8.4, dual combustion cycle) but it is the cycle to which

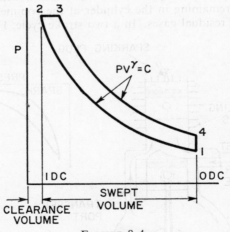

FIGURE 8.4

the older types of diesel engine, which used blast injection, approxi-
mated. The cycle is composed of four non-flow processes as follows:
At 1 with the piston at ODC, the cylinder is full of air. 1–2 represents
the reversible adiabatic compression of the air as the piston moves
from ODC to IDC. On the air standard cycle this compression process

would obey the equation $PV^\gamma = c$. At 2 the clearance volume of the cylinder contains hot high pressure air. As the piston starts to return to ODC, heat energy is supplied to the air from the surroundings resulting in a further increase of temperature of the air and the cylinder pressure remaining constant on this stroke until the supply of heat energy is cut off at 3. During the remainder of this stroke from 3 to 4 the air expands reversibly and adiabatically until the piston reaches ODC at 4. Between 4 and 1 heat energy flows from the air to the surroundings whilst the volume of the air remains constant, until the air regains its original condition at 1 ready for the cycle to begin again. Engines operating on this cycle may be either two- or four-stroke engines. Briefly, the main departures of actual engine Diesel cycles from the air standard Diesel cycle are:

(1) The actual diagram will be rounded at the junctions of the processes owing to the valve action not being instantaneous
(2) Compression of the air will not be reversible
(3) The supply of heat energy is obtained by the combustion of oil fuel which is blown into the cylinder (using compressed air) in the form of a spray at 2. Thus the mass of air in the cylinder is increased after 2
(4) The working fluid throughout the remainder of the cycle will not be air, but products of combustion
(5) The rejection of heat 4–1 is carried out by the removing of the products of combustion from the cylinder

To obtain an expression for the air standard efficiency of a Diesel cycle, consider the cylinder to contain 1 lb of air at 1.
In Fig. 8.4,

Let $\dfrac{V_1}{V_2}$ = compression ratio = r_v

$\dfrac{V_3}{V_2}$ = volume ratio of combustion = r_c

$$\text{Cycle efficiency} = \frac{\text{net work energy transferred}}{\text{heat energy supplied}}$$

$$= 1 - \frac{\text{heat energy rejected}}{\text{heat energy supplied}}$$

$$= 1 - \frac{Q_{41}}{Q_{23}} \tag{8.8}$$

Since 4–1 is a constant volume process

$$Q_{41} = c_v(T_4 - T_1) \text{ per lb} \tag{8.9}$$

Since 2–3 is a constant pressure process

$$Q_{23} = c_\text{p}(T_3 - T_2) \text{ per lb} \tag{8.10}$$

$$\therefore \quad \text{cycle efficiency} = 1 - \frac{c_\text{v}(T_4 - T_1)}{c_\text{p}(T_3 - T_2)}$$

$$= 1 - \frac{1}{\gamma}\frac{T_1}{T_2}\left\{\frac{(T_4/T_1)-1}{(T_3/T_2)-1}\right\} \tag{8.11}$$

For the constant pressure process 2–3 $\dfrac{V_2}{T_2} = \dfrac{V_3}{T_3}$

$$\therefore \quad \frac{T_3}{T_2} = \frac{V_3}{V_2} = r_\text{c} \tag{8.12}$$

For the isentropic process 1–2:

$$P_2 V_2{}^\gamma = P_1 V_1{}^\gamma \tag{8.13}$$

For the isentropic process 3–4

$$P_3 V_3{}^\gamma = P_4 V_4{}^\gamma \tag{8.14}$$

(8.13) divided by (8.14) gives

$$\frac{P_2 V_2{}^\gamma}{P_3 V_3{}^\gamma} = \frac{P_1 V_1{}^\gamma}{P_4 V_4{}^\gamma}$$

$$\therefore \quad \left(\frac{V_2}{V_3}\right)^\gamma = \frac{P_3}{P_2}\frac{P_1}{P_4}\left(\frac{V_1}{V_4}\right)^\gamma$$

Since $P_3 = P_2$ and $V_1 = V_4$ then

$$\left(\frac{V_2}{V_3}\right)^\gamma = \frac{P_1}{P_4} \tag{8.15}$$

For the constant volume process 4–1

$$\frac{P_4}{T_4} = \frac{P_1}{T_1}$$

$$\therefore \quad \frac{P_1}{P_4} = \frac{T_1}{T_4} \tag{8.16}$$

$$\therefore \quad \left(\frac{V_2}{V_3}\right)^\gamma = \frac{P_1}{P_4} = \frac{T_1}{T_4}$$

i.e. $$\frac{1}{r_\text{c}{}^\gamma} = \frac{T_1}{T_4}$$

or $$r_\text{c}{}^\gamma = \frac{T_4}{T_1} \tag{8.17}$$

Substituting in (8.11) gives

$$\text{cycle efficiency} = 1 - \frac{1}{\gamma}\frac{T_1}{T_2}\frac{(r_c{}^\gamma - 1)}{(r_c - 1)} \tag{8.18}$$

Since $P_1 V_1{}^\gamma = P_2 V_2{}^\gamma$ and $P_1 V_1 / T_1 = P_2 V_2 / T_2$

Then

$$\frac{T_2}{T_1} = \frac{P_2}{P_1}\frac{V_2}{V_1} = \left(\frac{V_1}{V_2}\right)^\gamma \left(\frac{V_1}{V_2}\right)^{-1} = \left(\frac{V_1}{V_2}\right)^{\gamma-1} = r_v{}^{\gamma-1} \tag{8.19}$$

$\therefore$

$$\text{cycle efficiency} = 1 - \frac{1}{\gamma}\frac{1}{r_v{}^{\gamma-1}}\frac{(r_c{}^\gamma - 1)}{(rc - 1)} \tag{8.20}$$

In the above expression, γ will be constant and for a certain engine r_v will be constant since it depends on the construction of the engine. Therefore, variation of cycle efficiency with engine load will depend on the variation of $(r_c{}^\gamma - 1)/(r_c - 1)$ with load. As more load is put on an engine, it necessitates the burning of more fuel, thus increasing r_c. Since both r_c and γ are greater than unity then as r_c increases, $(r_c{}^\gamma - 1)$ will increase faster than $(r_c - 1)$, resulting in $(r_c{}^\gamma - 1)/(r_c - 1)$ becoming greater as the engine load increases. Therefore, from the expression for cycle efficiency, an increase in load results in a decrease of efficiency. This factor combined with the need for additional equipment to provide the compressed air for fuel injection has caused the blast injection diesel engine to be mainly superseded by the solid injection diesel engine.

Ex. 8.2. An engine operating on the ideal Diesel cycle has a maximum pressure of 600 lbf/in² and a maximum temperature of 2,000°F. At the

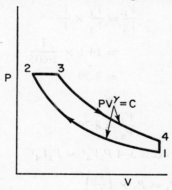

FIGURE E.8.2

beginning of the compression stroke the air is at 70°F and 14·7 lbf/in². Given that γ for air is equal to 1·4 determine the air standard efficiency of the cycle.

Since 1–2 is isentropic $P_1 V_1{}^\gamma = P_2 V_2{}^\gamma$

$$\therefore \quad \frac{V_1}{V_2} = \left(\frac{P_2}{P_1}\right)^{1/\gamma}$$

$$= \left(\frac{600 \text{ lbf/in}^2}{14 \cdot 7 \text{ lbf/in}^2}\right)^{1/1 \cdot 4}$$

$$= (40 \cdot 8)^{0 \cdot 715}$$

$$= 14 \cdot 1$$

Also $\quad \dfrac{P_1 V_1}{T_1} = \dfrac{P_2 V_2}{T_2}$

$$\therefore \quad T_2 = \frac{P_2}{P_1} \frac{V_2}{V_1} T_1$$

$$= \frac{600 \text{ lbf/in}^2}{14 \cdot 7 \text{ lbf/in}^2} \times \frac{1}{14 \cdot 1} \times (460 + 70) \text{ °R}$$

$$= 1,535 \text{°R}$$

Since $\quad \dfrac{P_2 V_2}{T_2} = \dfrac{P_3 V_3}{T_3}$

$$\frac{V_3}{V_2} = \frac{P_2}{P_3} \times \frac{T_3}{T_2}$$

$$= 1 \times \frac{2,460 \text{ °R}}{1,535 \text{ °R}}$$

$$= 1 \cdot 605$$

$$\therefore \quad \frac{V_1}{V_3} = \frac{V_1}{V_2} \times \frac{V_2}{V_3}$$

$$= 14 \cdot 1 \times \frac{1}{1 \cdot 605}$$

$$= 8 \cdot 79$$

Since $V_1 = V_4$ then

$$\frac{V_4}{V_3} = 8 \cdot 79$$

For the isentropic process 3–4 $P_3 V_3{}^\gamma = P_4 V_4{}^\gamma$

$$\therefore \quad P_4 = P_3 \times \left(\frac{V_3}{V_4}\right)^\gamma$$

$$= 600 \text{ lbf/in}^2 \times \left(\frac{1}{8 \cdot 79}\right)^{1 \cdot 4}$$

$$= 600 \text{ lbf/in}^2 \times \frac{1}{21} = 28 \cdot 6 \text{ lbf/in}^2$$

Since
$$\frac{P_1 V_1}{T_1} = \frac{P_4 V_4}{T_4}$$

$$T_4 = \frac{P_4}{P_1} \times \frac{V_4}{V_1} \times T_1$$

$$= \frac{28\cdot6 \text{ lbf/in}^2}{14\cdot7 \text{ lbf/in}^2} \times 1 \times 530 \text{ °R}$$

$$= 1{,}031\text{°R}$$

Considering 1 lb of air

heat energy supplied $= c_p \times (T_3 - T_2)$

$$= c_p \frac{\text{Btu}}{\text{lb degR}} \times (2{,}460 - 1{,}535) \text{ degR}$$

$$= 925 \, c_p \frac{\text{Btu}}{\text{lb}}$$

heat energy rejected $= c_v \times (T_4 - T_1)$

$$= c_v \frac{\text{Btu}}{\text{lb degR}} \times (1{,}031 - 530) \text{ degR}$$

$$= 501 \, c_v \frac{\text{Btu}}{\text{lb}}$$

$\therefore$ cycle efficiency $= 1 - \dfrac{\text{heat energy rejected}}{\text{heat energy supplied}}$

$$= 1 - \frac{501 \, c_v \, (\text{Btu/lb})}{925 \, c_p \, (\text{Btu/lb})}$$

$$= 1 - \frac{501}{925 \times \gamma}$$

$$= 1 - 0\cdot387 = 0\cdot613$$

8.4 Dual Combustion or Mixed Pressure Air Standard Cycle

This is the cycle to which that of a modern diesel engine approximates. It is a combination of the Otto and Diesel cycles and consists of five non-flow processes as follows:

At 1 the piston is at ODC and the cylinder is full of air. From 1–2 the air is compressed isentropically into the clearance volume, thus raising the pressure and temperature of the air. During 2–3 heat energy is supplied from the surroundings to further increase the pressure and temperature of the air, whilst the volume remains constant. At 3 the

pressure of the air reaches the maximum pressure attained in the
cycle. The pressure is maintained at this value from 3–4 whilst heat
energy is supplied from the surroundings, causing an increase in volume
and a further increase in temperature of the air. At 4 the temperature
of the air is the maximum value attained during the cycle. The air then
expands isentropically until the piston reaches ODC at 5. Heat energy
is then rejected to the surroundings by the air whilst the volume remains constant 5–1 until the air regains its original condition at 1
ready for the cycle to begin again.

As before, both two and four stroke engines may operate on this

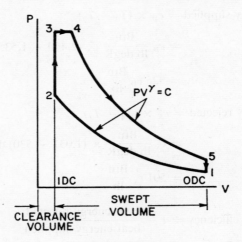

FIGURE 8.5

cycle. The differences between the cycle of an actual engine and the
air standard cycle are similar to those described for the Diesel cycle,
with the exception of the method of supplying the heat energy. In
this type of engine, the fuel is forced into the cylinder in the form
of a spray by the action of a fuel pump. Injection takes place both
before and after IDC.

When the fuel which has been injected into the cylinder before IDC
ignites, it burns almost instantaneously, so producing a constant
volume combustion process. The remainder of the fuel entering the
cylinder will then burn as soon as it leaves the nozzle, hence producing
a constant pressure combustion process. The actual mechanics of
these combustion processes are dealt with in more advanced courses in
applied thermodynamics.

An expression for the efficiency of the air standard mixed pressure
cycle may be obtained as follows (see Fig. 8.5)

Let $\dfrac{V_1}{V_2}$ = compression ratio = r_{v}

$\dfrac{P_3}{P_2}$ = pressure ratio of combustion = r_{p}

$\dfrac{V_4}{V_3}$ = volume ratio of combustion = r_{c}

Then cycle efficiency = $\dfrac{\text{net work energy transferred}}{\text{heat energy supplied}}$

$$= 1 - \dfrac{\text{heat energy rejected}}{\text{heat energy supplied}}$$

Considering the cylinder to contain 1 lb of fluid

Heat energy rejected per lb = Q_{51}
$$= c_{\mathrm{v}}(T_5 - T_1) \qquad (8.21)$$

Heat energy supplied per lb = $Q_{23} + Q_{34}$
$$= c_{\mathrm{v}}(T_3 - T_2) + c_{\mathrm{p}}(T_4 - T_3) \qquad (8.22)$$

$\therefore$ cycle efficiency = $1 - \dfrac{Q_{51}}{Q_{23} + Q_{34}}$

$$= 1 - \dfrac{c_{\mathrm{v}}(T_5 - T_1)}{c_{\mathrm{v}}(T_3 - T_2) + c_{\mathrm{p}}(T_4 - T_3)}$$

$$= 1 - \dfrac{c_{\mathrm{v}}T_1\{(T_5/T_1) - 1\}}{c_{\mathrm{v}}T_2\{(T_3/T_2) - 1\} + c_{\mathrm{p}}T_3\{(T_4/T_3) - 1\}}$$

$$= 1 - \dfrac{c_{\mathrm{v}}T_1\{(T_5/T_1) - 1\}}{c_{\mathrm{v}}T_2\{(T_3/T_2) - 1\} + c_{\mathrm{p}}T_2(T_3/T_2)\{(T_4/T_3)-1\}}$$
$$(8.23)$$

Since 5–1 is a constant volume process:

$$\dfrac{P_5}{T_5} = \dfrac{P_1}{T_1}$$

$\therefore$ $$\dfrac{T_5}{T_1} = \dfrac{P_5}{P_1} \qquad (8.24)$$

Since 1–2 and 4–5 are isentropic processes

$$P_1 V_1{}^\gamma = P_2 V_2{}^\gamma \text{ and } P_4 V_4{}^\gamma = P_5 V_5{}^\gamma$$

$$\therefore \qquad \frac{T_5}{T_1} = \frac{P_5}{P_1}$$

$$= \frac{P_4 \, (V_4/V_5)^\gamma}{P_2 \, (V_2/V_1)^\gamma}$$

$$= \frac{P_4}{P_2} \left(\frac{V_4}{V_2}\right)^\gamma \left(\frac{V_1}{V_5}\right)^\gamma$$

$$= \frac{P_4}{P_2} \left(\frac{V_4}{V_2}\right)^\gamma \text{ since } V_1 = V_5$$

$$= \frac{P_4}{P_2} \left(\frac{V_4}{V_3}\right)^\gamma \text{ since } V_2 = V_3$$

$$= \frac{P_3}{P_2} \left(\frac{V_4}{V_3}\right)^\gamma \text{ since } P_4 = P_3$$

i.e.
$$\frac{T_5}{T_1} = r_{\mathrm p} \, r_{\mathrm c}{}^\gamma \qquad\qquad (8.25)$$

Since 2–3 is a constant volume process

$$\frac{P_3}{T_3} = \frac{P_2}{T_2}$$

i.e.
$$\frac{T_3}{T_2} = \frac{P_3}{P_2} = r_{\mathrm p} \qquad\qquad (8.26)$$

Since 3–4 is a constant pressure process

$$\frac{V_3}{T_3} = \frac{V_4}{T_4}$$

i.e.
$$\frac{T_4}{T_3} = \frac{V_4}{V_3} = r_{\mathrm c} \qquad\qquad (8.27)$$

Substituting from (8.25), (8.26) and (8.27) in (8.23) gives

$$\text{cycle efficiency} = 1 - \frac{c_{\mathrm v} T_1 (r_{\mathrm p} r_{\mathrm c}{}^\gamma - 1)}{c_{\mathrm v} T_2 (r_{\mathrm p} - 1) + c_{\mathrm p} T_2 r_{\mathrm p} (r_{\mathrm c} - 1)}$$

$$= 1 - \frac{T_1}{T_2}\left[\frac{r_{\mathrm p} r_{\mathrm c}{}^\gamma - 1}{(r_{\mathrm p} - 1) + \gamma r_{\mathrm p} (r_{\mathrm c} - 1)} \right] \qquad (8.28)$$

Since 1–2 is an isentropic process

$$P_1 V_1{}^\gamma = P_2 V_2{}^\gamma \qquad\qquad (8.29)$$

Also
$$\frac{P_1 V_1}{T_1} = \frac{P_2 V_2}{T_2}$$

$$\therefore \qquad \frac{T_2}{T_1} = \frac{P_2 V_2}{P_1 V_1}$$

$$= \left(\frac{V_1}{V_2}\right)^{\gamma} \left(\frac{V_1}{V_2}\right)^{-1} \qquad \text{from (8.29)}$$

$$= \left(\frac{V_1}{V_2}\right)^{\gamma-1}$$

$$= r_{\mathrm{v}}{}^{\gamma-1}$$

Substituting in (8.28) gives

$$\text{cycle efficiency} = 1 - \frac{1}{r_{\mathrm{v}}{}^{\gamma-1}}\left[\frac{r_{\mathrm{p}} r_{\mathrm{c}}{}^{\gamma} - 1}{(r_{\mathrm{p}} - 1) + \gamma r_{\mathrm{p}}(r_{\mathrm{c}} - 1)}\right] \qquad (8.30)$$

It should be noted that the more heat that is supplied at constant volume, the greater will be the maximum pressure and the greater the efficiency.

Ex. 8.3. An oil engine operating on the ideal mixed pressure cycle at a compression ratio of 16:1, takes in air at 14·7 lbf/in² and 70°F.

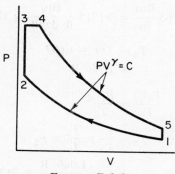

FIGURE E.8.3

During the cycle the heat energy supplied per lb of air is 65 Btu at constant volume and 120 Btu at constant pressure. Given that for air $c_{\mathrm{p}} = 0.24$ Btu/lb deg R and $c_{\mathrm{v}} = 0.1715$ Btu/lb degR determine:

(a) The maximum pressure and temperature of the cycle
(b) The work energy transferred per lb of air
(c) The cycle efficiency

$$\gamma = \frac{c_{\mathrm{p}}}{c_{\mathrm{v}}} = \frac{0.24}{0.1715} = 1.4$$

Since 1–2 is isentropic $P_1V_1{}^\gamma = P_2V_2{}^\gamma$

$$\therefore \qquad P_2 = P_1 \left(\frac{V_1}{V_2}\right)^\gamma$$
$$= 14{\cdot}7 \ (\mathrm{lbf/in^2}) \times (16)^{1{\cdot}4}$$
$$= 14{\cdot}7 \ (\mathrm{lbf/in^2}) \times 48{\cdot}5$$
$$= 712 \ \mathrm{lbf/in^2}$$

Since $\qquad \dfrac{P_1V_1}{T_1} = \dfrac{P_2V_2}{T_2}$

$$T_2 = \frac{P_2}{P_1} \times \frac{V_2}{V_1} \times T_1$$
$$= \frac{712 \ \mathrm{lbf/in^2}}{14{\cdot}7 \ \mathrm{lbf/in^2}} \times \frac{1}{16} \times (460 + 70) \ ^\circ\mathrm{R}$$
$$= 1{,}607 ^\circ\mathrm{R}$$

For constant volume process 2–3

$$\text{heat energy supplied} = mc_\mathrm{v}(T_3 - T_2)$$

i.e. per lb of air $\qquad 65 \ \dfrac{\mathrm{Btu}}{\mathrm{lb}} = 0{\cdot}1715 \ \dfrac{\mathrm{Btu}}{\mathrm{lb \ degR}} \times (T_3 - 1{,}607) \ \mathrm{degR}$

$$\therefore \qquad T_3 = 379 + 1{,}607$$
$$= 1{,}986 ^\circ\mathrm{R}$$

Since $\qquad \dfrac{P_2V_2}{T_2} = \dfrac{P_3V_3}{T_3}$

$$P_3 = \frac{V_2}{V_3} \times \frac{T_3}{T_2} \times P_2$$
$$= 1 \times \frac{1{,}986 \ ^\circ\mathrm{R}}{1{,}607 \ ^\circ\mathrm{R}} \times 712 \ \mathrm{lbf/in^2}$$
$$= 880 \ \mathrm{lbf/in^2}$$

For constant pressure process 3–4

$$\text{heat energy supplied} = mc_\mathrm{p}(T_4 - T_3)$$

i.e. per lb of air $\qquad 120 \ \dfrac{\mathrm{Btu}}{\mathrm{lb}} = 0{\cdot}24 \ \dfrac{\mathrm{Btu}}{\mathrm{lb \ degR}} \times (T_4 - 1{,}986) \ \mathrm{degR}$

$$\therefore \qquad T_4 = 500 + 1{,}986$$
$$= 2{,}486 ^\circ\mathrm{R}$$

Since
$$\frac{P_3 V_3}{T_3} = \frac{P_4 V_4}{T_4}$$

$$V_4 = \frac{P_3}{P_4} \times \frac{T_4}{T_3} \times V_3$$

$$= 1 \times \frac{2{,}486\ ^\circ R}{1{,}986\ ^\circ R} \times V_2$$

$$= 1\cdot252\ V_2$$

Since 4–5 is isentropic $P_4 V_4{}^\gamma = P_5 V_5{}^\gamma$

$$\therefore \qquad P_5 = \left(\frac{V_4}{V_5}\right)^\gamma \times P_4$$

$$= \left(\frac{1\cdot252\ V_2}{V_1}\right)^\gamma \times 880\ \text{lbf/in}^2$$

$$= \left(1\cdot252 \times \frac{1}{16}\right)^{1\cdot4} \times 880\ \text{lbf/in}^2$$

$$= 0\cdot0284 \times 880\ \text{lbf/in}^2$$

$$= 250\ \text{lbf/in}^2$$

Since
$$\frac{P_5 V_5}{T_5} = \frac{P_1 V_1}{T_1}$$

$$T_5 = \frac{P_5}{P_1} \times \frac{V_5}{V_1} \times T_1$$

$$= \frac{250\ \text{lbf/in}^2}{14\cdot7\ \text{lbf/in}^2} \times 1 \times 530\ ^\circ R$$

$$= 900\ ^\circ R$$

Total heat energy supplied per lb = 65 Btu/lb + 120 Btu/lb
$$= 185\ \text{Btu/lb}$$

heat energy rejected per lb = $c_v(T_5 - T_1)$

$$= 0\cdot1715\ \frac{\text{Btu}}{\text{lb degR}} \times (900 - 530)\ \text{degR}$$

$$= 63\cdot5\ \frac{\text{Btu}}{\text{lb}}$$

$$\therefore \qquad \text{cycle efficiency} = 1 - \frac{\text{heat energy rejected}}{\text{heat energy supplied}}$$

$$= 1 - \frac{63\cdot5\ \text{Btu/lb}}{185\ \text{Btu/lb}}$$

$$= 1 - 0\cdot343$$

$$= 0\cdot657$$

8.5 Standard Cycle for Gas Turbine Plant

The fundamental difference between the air standard cycle for gas turbine plant and the cycle on which reciprocating engines operate is that in a gas turbine plant all the processes are flow processes. In the simple treatment of the cycle, all the processes are considered to be steady flow processes, and changes in kinetic and potential energy in the various components of the plant are neglected. The air standard cycle, which is known as the constant pressure cycle or Joule cycle, is

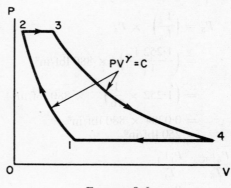

FIGURE 8.6

composed of four processes, which may be described on a $P–V$ diagram as shown in Fig. 8.6.

1–2. Air at 1 is compressed reversibly and adiabatically, thus following the equation $PV^\gamma = C$, to a higher pressure and temperature at 2.

2–3. Heat energy is then supplied to the air from the surroundings whilst the pressure of the air remains constant. This heat energy causes the temperature of the air to rise to the maximum temperature attained in the cycle at 3.

3–4. The air is expanded adiabatically and reversibly, thus following the equation $PV^\gamma = C$, until the pressure of the air reaches its original value at 4.

4–1. Heat energy is transferred from the air to the surroundings whilst the pressure of the air remains constant to close the cycle.

A line diagram of a simple gas turbine plant is shown in Fig. 8.7 and numbered so as to correspond with the $P–V$ diagram shown above. Consider a flow of m lb of air flowing round the plant. Air enters a rotary compressor at state 1 and is compressed isentropically to state 2. Applying the steady flow energy equation to the

compressor and neglecting changes in kinetic energy and potential
energy gives

$$- W_{12}/\text{J} = H_2 - H_1$$
$$= m(h_2 - h_1)$$

or $$\qquad W_{12}/\text{J} = m(h_1 - h_2) \qquad (8.31)$$

The air then enters a heater, where heat energy Q_{23} is transferred to the
air from the surroundings. Applying the steady flow energy equation

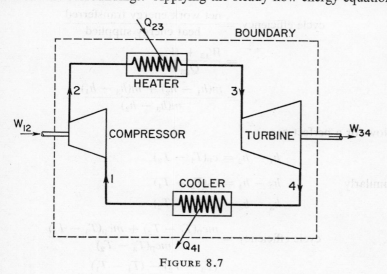

FIGURE 8.7

to the heater and neglecting changes in kinetic and potential energy
gives

$$Q_{23} = H_3 - H_2$$
$$= m(h_3 - h_2) \qquad (8.32)$$

After leaving the heater, the air is expanded isentropically in a turbine,
during which processes work energy W_{34} is transferred from the air to
the surroundings. Applying the steady flow energy equation to the
turbine and neglecting the change in kinetic and potential energy gives

$$- W_{34} = H_4 - H_3$$
$$= m(h_4 - h_3)$$

or $$\qquad W_{34} = m(h_3 - h_4) \qquad (8.33)$$

The air is then cooled to its original state in a cooler where heat energy
Q_{41} is transferred from the air to the surroundings. Applying the steady

flow energy equation to the cooler and neglecting changes in kinetic and potential energy gives

$$Q_{41} = H_1 - H_4$$
$$= m(h_1 - h_4) \qquad (8.34)$$

An expression for the efficiency of the air standard cycle may be obtained as follows:

$$\text{cycle efficiency} = \frac{\text{net work energy transferred}}{\text{heat energy supplied}}$$

$$= \frac{W_{12} + W_{34}}{Q_{23}}$$

$$= \frac{m(h_1 - h_2) + m(h_3 - h_4)}{m(h_3 - h_2)}$$

Now for a perfect gas,

$$h_1 - h_2 = c_p(T_1 - T_2)$$

Similarly
$$h_3 - h_4 = c_p(T_3 - T_4)$$
$$h_3 - h_2 = c_p(T_3 - T_2)$$

$$\therefore \qquad \text{cycle efficiency} = \frac{mc_p(T_1 - T_2) + mc_p(T_3 - T_4)}{mc_p(T_3 - T_2)}$$

$$= \frac{(T_3 - T_2) - (T_4 - T_1)}{(T_3 - T_2)}$$

$$= 1 - \frac{T_4 - T_1}{T_3 - T_2}$$

$$= 1 - \frac{T_1\{(T_4/T_1) - 1\}}{T_2\{(T_3/T_2) - 1\}} \qquad (8.35)$$

It should be confirmed by the student that this expression can also be obtained using the expression

$$\text{cycle efficiency} = 1 - \frac{\text{heat rejected}}{\text{heat supplied}}$$

Now since 1–2 is an isentropic process, $P_1 V_1{}^\gamma = P_2 V_2{}^\gamma$

also
$$\frac{P_1 V_1}{T_1} = \frac{P_2 V_2}{T_2}$$

$$\therefore \quad \frac{T_2}{T_1} = \frac{P_2}{P_1}\frac{V_2}{V_1}$$

$$= \frac{P_2}{P_1}\left(\frac{P_1}{P_2}\right)^{1/\gamma}$$

$$= \left(\frac{P_2}{P_1}\right)^{1}\left(\frac{P_2}{P_1}\right)^{-1/\gamma}$$

$$= \left(\frac{P_2}{P_1}\right)^{(\gamma-1)/\gamma} \quad (8.36)$$

Similarly since 3–4 is an isentropic process $P_3V_3{}^\gamma = P_4V_4{}^\gamma$

also
$$\frac{P_3V_3}{T_3} = \frac{P_4V_4}{T_4}$$

$$\therefore \quad \frac{T_3}{T_4} = \frac{P_3}{P_4}\frac{V_3}{V_4} = \frac{P_3}{P_4}\left(\frac{P_4}{P_3}\right)^{1/\gamma} = \left(\frac{P_3}{P_4}\right)^{1}\left(\frac{P_3}{P_4}\right)^{-1/\gamma} = \left(\frac{P_3}{P_4}\right)^{(\gamma-1)/\gamma} \quad (8.37)$$

But $P_2 = P_3$ and $P_1 = P_4$

$$\therefore \quad \left(\frac{P_2}{P_1}\right)^{(\gamma-1)/\gamma} = \left(\frac{P_3}{P_4}\right)^{(\gamma-1)/\gamma}$$

$$\therefore \quad \frac{T_2}{T_1} = \frac{T_3}{T_4}$$

$$\therefore \quad \frac{T_4}{T_1} = \frac{T_3}{T_2}$$

$$\therefore \quad \frac{T_4}{T_1} - 1 = \frac{T_3}{T_2} - 1 \quad (8.38)$$

$$\therefore \quad \text{cycle efficiency} = 1 - \frac{T_1}{T_2} = 1 - \frac{1}{(P_2/P_1)^{(\gamma-1)/\gamma}} = 1 - \frac{1}{r_{\mathrm{p}}^{(\gamma-1)/\gamma}}$$

$$(8.39)$$

where r_{p} is the pressure ratio $= P_2/P_1$

This efficiency will be lower than that of a Carnot cycle operating between the same temperature limits since T_2 is not the maximum cycle temperature.

The simplest gas turbine operates on an open cycle. In this cycle, atmospheric air is drawn into the compressor where it undergoes an irreversible adiabatic compression process. The heater is replaced by a combustion chamber where fuel is burned continuously in the high pressure air, and the expansion of the products of combustion in the turbine is adiabatic but irreversible. In addition to being coupled to the external load the turbine is also coupled to the compressor to provide the work energy necessary to drive the compressor. A cooler is not included in the circuit, the products of combustion being expelled into the atmosphere from the turbine exhaust. The maximum temperature attained in this cycle is of the utmost importance, since the combustion chamber is subject to this temperature continuously, and not intermittently, as when the combustion takes place in a reciprocating engine. Hence the maximum temperature of a gas turbine cycle is generally much lower than the maximum temperature attained in a reciprocating engine cycle.

Ex. 8.4. A simple open cycle gas turbine plant operates with a pressure ratio of 4:1 over a temperature range from 65°F to 1,300°F. Determine the thermal efficiency of the plant, and the rate of air flow

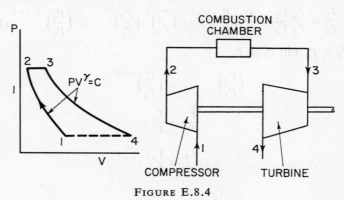

FIGURE E.8.4

required for a net theoretical turbine output of 500 horsepower. Take $\gamma = 1.4$ throughout.

Since 1–2 is isentropic $P_1V_1^\gamma = P_2V_2^\gamma$

Also
$$\frac{P_1V_1}{T_1} = \frac{P_2V_2}{T_2}$$

$\therefore$
$$\frac{T_2}{T_1} = \frac{P_2}{P_1}\frac{V_2}{V_1}$$

$$= \frac{P_2}{P_1}\left(\frac{P_1}{P_2}\right)^{1/\gamma}$$

$$= \left(\frac{P_2}{P_1}\right)^{(\gamma-1)/\gamma} = 4^{(1.4-1)/1.4} = 1.487$$

(Similarly, since 3–4 is isentropic, $\dfrac{T_3}{T_4} = \left(\dfrac{P_3}{P_4}\right)^{(\gamma-1)/\gamma} = 4^{(1\cdot4-1)/1\cdot4} = 1\cdot487$

$$\therefore\ T_2 = 1\cdot487 \times T_1$$
$$= 1\cdot487 \times (460 + 65)\,°R$$
$$= 780°R$$

and $\qquad T_4 = \dfrac{T_3}{1\cdot487}$

$$= \dfrac{(1{,}300 + 460)°R}{1\cdot487}$$

$$= 1{,}185°R$$

Considering 1 lb of air

Heat energy supplied per lb $= c_p(T_3 - T_2)$

$$= 0\cdot24\ \dfrac{\text{Btu}}{\text{lb degR}} \times (1{,}760 - 780)\,\text{degR}$$

$$= 235\ \text{Btu/lb}$$

Heat energy rejected per lb $= c_p(T_4 - T_1)$

$$= 0\cdot24\ \dfrac{\text{Btu}}{\text{lb degR}} \times (1{,}185 - 525)\,\text{degR}$$

$$= 158\ \text{Btu/lb}$$

$\therefore$ Net work energy transferred $=$ heat energy supplied $-$ heat energy rejected

$$= 235 - 158 = 77\ \text{Btu/lb}$$

Cycle efficiency $= \dfrac{\text{net work energy transferred}}{\text{heat energy supplied}}$

$$= \dfrac{77\ \text{Btu/lb}}{235\ \text{Btu/lb}} = 0\cdot328$$

500 horsepower $= 500 \times 33{,}000\ \text{ft lbf/min}$

$$= \dfrac{500 \times 33{,}000}{778}\ \dfrac{\text{ft lbf/min}}{\text{ft lbf/Btu}}$$

$$= 21{,}200\ \text{Btu/min}$$

rate of flow $= \dfrac{\text{energy required}}{\text{energy transferred per lb}}$

$$= \dfrac{21{,}2000\ \text{Btu/min}}{77\ \text{Btu/lb}} = 276\ \text{lb/min}$$

EXERCISES ON CHAPTER 8

N.B. All pressures are absolute. For air, take $c_p = 0.238$ Btu/lb degR and $c_v = 0.17$ Btu/lb degR.

1. During an ideal Otto cycle, 1 lb of air is compressed from 14·7 lbf/in² and 60°F to a pressure of 220 lbf/in². The heat energy supplied at constant volume increases the temperature to 3,500°F. Calculate the efficiency of the cycle and the mean effective pressure.

$$(53.8\%; \ 126 \ \text{lbf/in}^2)$$

2. An engine operating on the ideal Otto cycle has a compression ratio of 7·5:1. At the beginning of the compression stroke the air is at 15 lbf/in² and 80°F. If the heat energy supplied is 400 Btu per lb of air, calculate the maximum temperature and pressure in the cycle, and the cycle efficiency.

$$(3,562°\text{R}; \ 743 \ \text{lbf/in}^2; \ 55.25\%)$$

3. A petrol engine has a bore of 3·5 in, a stroke of 4 in, and a clearance volume of 5·7 in³. If the thermal efficiency of the engine is 28%, calculate the relative efficiency of the engine.

$$(50.1\%)$$

4. An engine with a compression ratio of 15:1 operates on the ideal Diesel cycle and takes in air at 14·7 lbf/in² and 80°F. If the temperature at the end of combustion is 3,000°F, calculate the pressure at which combustion takes place, the percentage of the stroke at which combustion is complete, and the thermal efficiency of the cycle.

$$(650 \ \text{lbf/in}^2; \ 8.4\%; \ 59.4\%)$$

5. At the beginning of compression, the charge in the cylinder of an engine operating on the ideal Diesel cycle is at 15 lbf/in² and 90°F, and occupies a volume of 9 ft³. After adiabatic compression through a compression ratio of 13:1, 300 Btu of heat energy are supplied at constant pressure. Calculate the maximum temperature and pressure attained in the cycle, the cycle efficiency, and the mean effective pressure.

$$(3,415°\text{R}; \ 543 \ \text{lbf/in}^2; \ 57.1\%; \ 111.5 \ \text{lbf/in}^2)$$

6. In an engine operating on the ideal mixed pressure cycle, the compression ratio is 12·5:1. Initially the cylinder contents are at 14·7 lbf/in² and 80°F. If the maximum pressure in the cycle is 850 lbf/in² and the maximum temperature is 3,500°F, calculate the cycle efficiency and the mean effective pressure.

$$(61\%; \ 135 \ \text{lbf/in}^2)$$

7. An engine having a compression ratio of 15:1 operates on the ideal mixed pressure cycle. At the beginning of compression the air in the cylinder is at 14·3 lbf/in² and 70°F. After compression, the heat energy supplied at constant volume increases the temperature to 2,000°F, and the heat energy supplied at constant pressure takes place over 4% of the stroke. Calculate the maximum pressure, the maximum temperature, and the cycle efficiency.

$$(996 \ \text{lbf/in}^2; \ 3,840°\text{R}; \ 63.8\%)$$

8. A simple open cycle gas turbine operates with a pressure ratio of 4·5:1. If the maximum and minimum cycle temperatures are 1,200°F and 80°F respectively, calculate for a flow rate of 1 lb/s:

(*a*) The rate of fuel consumption (in lb/h) if the calorific value of the fuel is 18,000 Btu/lb

(b) The horse power available for driving an external load
(c) The plant efficiency

(39·8 lb/h; 97·6 hp; 35%)

9. The compressor of a simple open cycle gas turbine plant raises the temperature of the air from 70°F to 300°F. The gases leave the combustion chamber and enter the turbine at a temperature of 1,250°F. If fuel of calorific value 19,000 Btu/lb is used in the combustion chamber at the rate of 10 lb/min, calculate the pressure ratio of the plant, the turbine exhaust temperature, the turbine horsepower, the horsepower absorbed by the compressor and the plant efficiency.

(3·5:1; 735°F; 2,430 hp; 1,082 hp; 30%)

10. A theoretical air cycle consists of three non-flow processes:

(i) Heat energy supplied at constant pressure to raise the temperature of the air from T_1°R to T_2°R
(ii) Reversible adiabatic expansion until the temperature reaches T_1°R
(iii) Heat energy rejected during isothermal compression at T_1°R to close the cycle

Show that the cycle efficiency is given by the expression

$$1 - \frac{T_1}{T_2 - T_1} \ln \frac{T_2}{T_1}$$

9: Combustion

Combustion is a rapid chemical reaction during which the elements in a fuel combine with the oxygen in the air thereby releasing heat energy. This energy is the source of heat for heat engines and, therefore, some knowledge of the chemistry of combustion is very important in the study of applied thermodynamics.

9.1 The Chemistry of Combustion

Most students will have studied chemistry at some time but those students who have not may be assured that the knowledge of chemistry required for an understanding of this chapter is elementary and easily mastered.

All substances present in the universe are elements or combinations of two or more elements. Altogether just over one hundred elements have been discovered. Elements themselves are the basic 'building bricks' of the universe; they cannot be sub-divided into more elementary constituents. Each of the elements are represented by a shorthand system in which capital letters denote the elements. Thus, carbon is represented by C, hydrogen by H, oxygen by O, nitrogen by N, etc. All elements are made up of *atoms*. The smallest unit, however, of any element which can exist by itself in a stable form is the *molecule* which is made up of one or more atoms. A monatomic element is one in which the molecule contains one atom only; a diatomic element is one in which the molecule is made up of two atoms; the element is triatomic when the molecule has three atoms and polyatomic when the molecule has more than three atoms.

When elements combine chemically new substances are formed called *compounds*. These compounds generally have vastly different properties from the elements out of which they have been formed.

The shorthand system discussed above can also be made to show the number of atoms in the molecule. For instance, O_2 shows that the element oxygen has two atoms in the molecule, C shows that carbon is a monatomic element having one atom per molecule. It will be noticed that the suffix 1 is omitted in this case. Similarly, H_2 and N_2 shows that hydrogen and nitrogen are diatomic molecules, whereas S shows that sulphur is monatomic.

9.2 Atomic Weight

Each atom of an element has a mass which is subjected to the gravitational force, i.e. it possesses weight. This weight is constant for the atom of a particular element. The weight of an individual atom is so small that even the most sensitive balance would not even begin to detect it. It was, therefore, decided that one particular element should be taken as a standard and all others related to it. Oxygen was taken as a standard and one atom of oxygen was given the atomic weight of 16. The atomic weight of the lightest element, hydrogen (to the nearest whole number) is 1. The atomic weights of the elements commonly found in combustion processes are:

Carbon, 12; Sulphur, 32; Oxygen, 16; Hydrogen, 1; Nitrogen, 14.

Molecular weights

Molecular weight is the weight of each molecule relative to a standard atomic weight of oxygen.

For monatomic elements, the molecular weight is equal to the atomic weight.

For diatomic elements the molecular weight is twice the atomic weight.

Oxygen, hydrogen and nitrogen exists as diatomic molecules having molecular weights of 32, 2 and 28 respectively.

The molecular weights of the monatomic elements of carbon and sulphur are 12 and 32 respectively.

The molecular weight of the gas carbon dioxide (CO_2) is $12 + 32 = 44$, and of steam (H_2O) is $2 + 16 = 18$.

9.3 Chemical Equations of Combustion

As previously stated combustion is the result of the combination of the elements in the fuel with oxygen. If carbon is burned with an adequate supply of oxygen, carbon dioxide gas is formed. This is represented by the equation:

$$C + O_2 \rightarrow CO_2 \qquad (9.1)$$

i.e. 1 molecule of carbon (solid) + 1 molecule of oxygen produces 1 molecule of carbon dioxide.

The number of atoms on the L.H.S. = 1 atom of carbon + 2 atoms of oxygen = 3. On the R.H.S. it is also equal to 3. It must be noted that on both sides of the equation the number of atoms of each element must balance. This rule assists in determining the correct equation for any reaction, as will be seen later.

Consider the burning of hydrogen. Here the equation may be written

$$H_2 + xO_2 \rightarrow H_2O \text{ (water vapour or water)} \qquad (9.2)$$

The number of H_2 atoms on the L.H.S. and R.H.S. must be the same. This condition must be satisfied in equation (9.2).

For the oxygen, we have $2x$ atoms on the L.H.S. and 1 atom on the R.H.S.

i.e. $\qquad\qquad\qquad 2x = 1$
or $\qquad\qquad\qquad\quad x = \frac{1}{2}$

The equation may now be written in the form

$$H_2 + \tfrac{1}{2}O_2 \rightarrow H_2O \qquad (9.3)$$

Since it is not possible to have one half of a molecule of any substance, both sides of the equation are multiplied by 2 to give:

$$2H_2 + O_2 \rightarrow 2H_2O \qquad (9.4)$$

If the carbon is burned with an insufficient supply of oxygen, carbon monoxide gas is formed. This is a very dangerous, toxic gas which is likely to result in death if inhaled in enclosed spaces. The exhaust of a motor car engine contains carbon monoxide and for this reason it is unwise to work in a closed garage with the car engine running. Its presence, also, denotes inefficient combustion, as only a portion of the carbon is burned completely to liberate heat energy. The equation can be obtained as follows:

$$C + xO_2 \rightarrow CO \qquad (9.5)$$

Here, the oxygen balance gives

$$2x = 1$$
$$x = \tfrac{1}{2}$$

$\therefore \qquad\qquad\qquad C + \tfrac{1}{2}O_2 \rightarrow CO \qquad (9.6)$
or $\qquad\qquad\qquad 2C + O_2 \rightarrow 2CO \qquad (9.7)$

The combustion of sulphur is represented by the equation:

$$\underset{\text{sulphur}}{S} + O_2 \rightarrow \underset{\substack{\text{sulphur} \\ \text{dioxide}}}{SO_2} \qquad (9.8)$$

Ex. 9.1 Form the chemical equation representing the complete combustion of (i) CH_4, (ii) C_7H_{16}.

The complete combustion of a hydrocarbon compound will always result in the formation of carbon dioxide and water.

(i) The equation for combustion of CH_4 (methane gas) can be written as follows:

$$CH_4 + xO_2 \rightarrow CO_2 + yH_2O$$

Hydrogen balance

$$4 \text{ atoms} = 2y \text{ atoms}$$
$$y = 2$$

Oxygen balance

$$2x = 2 + y$$
$$2x = 2 + 2$$
$$x = 4/2 = 2$$

The equation is given by

$$CH_4 + 2O_2 \rightarrow CO_2 + 2H_2O$$

(ii) The equation for combustion of C_7H_{16} can be written as follows:

$$C_7H_{16} + xO_2 \rightarrow 7CO_2 + 8H_2O$$

It can be seen by inspection that 7 molecules of CO_2 and 8 molecules of H_2O can be formed from the number of atoms present on the L.H.S.

Oxygen balance

$$2x \text{ atoms} = 14 + 8 = 22$$
$$x = 22/2 = 11$$

The equation of combustion is given by:

$$C_7H_{16} + 11O_2 \rightarrow 7CO_2 + 8H_2O$$

9.4 Analysis by Weight and Volume

It was seen from the foregoing articles that the reactions followed definite laws and produced definite compounds. The weights and volumes of the products and the reactants in a chemical reaction can easily be obtained from the equation. Consider the equation for the complete combustion of carbon, i.e.

$$C \quad + O_2 \rightarrow \quad CO_2 \qquad (9.9)$$

1 molecule + 1 molecule produces 2 molecules

Molecular weight   12   32   44

It is obvious that this may be written as

$$12 \text{ lb} + 32 \text{ lb} \qquad 44 \text{ lb}$$

or

$$1 \text{ lb} \qquad \frac{32}{12} \text{ lb} \qquad \frac{44}{12} \text{ lb}$$

In order to determine the volume of the reactants and products resort is made to Avogadro's Law, which states that equal volumes of different gases at the same temperature and pressure contain the same number of molecules.

A quantity of gas which has a weight equal to its molecular weight in pounds is known as a pound mole (usually written as lb mol or simply a mol). Thus, 1 mol of oxygen (molecular weight 32) weighs 32 lbf, 1 mol of hydrogen (molecular weight 2), weighs 2 lbf, etc. Now it follows from Avogadro's Law that equal volumes of different gases at the same temperature and pressure will have heights in direct proportion to the molecular weight (since the number of molecules will be the same for each gas). It follows, also, that if several gases are considered at the same temperature and pressure, such that the weight of each gas is equal to its molecular weight in pounds then the volume that they will occupy will be the same in each case. This particular volume has been found to be 359 ft³ at 14·7 lbf/in² and 32°F. Thus, at the given pressure and temperature, 32 lb of oxygen, 2 lb of hydrogen, 28 lb of nitrogen, etc., will all occupy the same volume of 359 ft³. Consider 1 mol of a gas which has a molecular weight of X. Applying the characteristic gas equation gives

$$PV_x = XR_xT \text{ or } \frac{PV_x}{T} = XR_x$$

Similarly, for 1 mol of a gas of molecular weight Y

$$PV_y = YR_yT \text{ or } \frac{PV_y}{T} = YR_y$$

Now it has been shown that the volume occupied by a mol of any gas under the same pressure and temperature conditions is the same, i.e. $V_x = V_y$.

Hence $\dfrac{PV_x}{T} = \dfrac{PY_y}{T}$

or $\qquad XR_x = YR_y = \text{constant } R_0$

The constant R_0 is known as the Universal Gas Constant and has the same value for all gases.

The above equation may be written

$$PV_m = R_0T \text{ where } V_m \text{ is the volume of a mol of gas}$$

If the value of the volume of 1 lb mol of 359 ft³ at 14·7 lbf/in² and 32°F is substituted in this equation the value of R_0 may be obtained

$$(14·7 \times 144) \text{ lbf/ft}^2 \times 359 \text{ ft}^3/\text{mol} = R_0 \times 492°R$$

∴ $\qquad R_0 = 1,545 \text{ ft lbf/mol degR}$

The characteristic constant for any gas may be obtained by dividing R_0 by the molecular weight of the gas, e.g.

$$\text{for oxygen } R = \frac{1,545}{32} = 48\cdot25 \text{ ft lbf/lb degR}$$

$$\text{for hydrogen } R = \frac{1,545}{2} = 772\cdot5 \text{ ft lbf/lb degR}$$

It will be appreciated that the volume of 1 mol of a solid will be negligible when compared with 359 ft³ and may be neglected.

Hence the volumes in equation (9.9) may be found as follows:

	C	+	O_2	→	CO_2	(9.10)
Mass equation	12 lb		32 lb		44 lb	
'Mol' equation	1 mol		1 mol		1 mol	
Volume equation	0 ft³		359 ft³		359 ft³	at S.T.P.
or	0 ft³		1 ft³		1 ft³	

For equation (9.7) the corresponding equations are:

	2C	+	O_2	→	2CO	(9.11)
Mass equation	24 lb		32 lb		56 lb	
'Mol' equation	2 mol		1 mol		2 mol	
Volume equation	0 ft³		359 ft³		2 × 359 ft³	at S.T.P.
or	0 ft³		1 ft³		2 ft³	

For the combustion of hydrogen the equations are:

	H_2	+	O_2	→	$2H_2O$	(9.12)
Mass equation	4 lb		32 lb		36 lb	
'Mol' equation	2 mol		1 mol		2 mol	
Volume equation	2 × 359 ft³		359 ft³		2 × 359 ft³	at S.T.P.
or	2 ft³		1 ft³		2 ft³	(if in the form of a vapour)
or	2 ft³		1 ft³		0 ft³	(if in the form of a liquid)

For the combustion of sulphur the equations are:

	S	+	O_2	→	SO_2	
Mass equation	32 lb		32 lb		64 lb	
'Mol' equation	1 mol		1 mol		1 mol	
Volume equation	0 ft³		359 ft³		359 ft³	
or	0 ft³		1 ft³		1 ft³	

Notice that the volumes on L.H.S. and R.H.S. are not required to balance.

A knowledge of Avogadro's Law enables a determination of the specific volumes of gases to be made.

Ex. 9.2. Determine the specific volume of oxygen, hydrogen and methane at the standard pressure and temperature of 14·7 lbf/in² and 32°F.

Specific volume of oxygen $= \dfrac{359 \text{ ft}^3}{32 \text{ lb}} = 11\cdot21 \text{ ft}^3/\text{lb}$

Specific volume hydrogen $= \dfrac{359 \text{ ft}^3}{2 \text{ lb}} = 179\cdot5 \text{ ft}^3/\text{lb}$

Specific volume methane $(CH_4) = \dfrac{359 \text{ ft}^3}{(12 + 4)\text{lb}} = 22\cdot4 \text{ ft}^3/\text{lb}$

Ex. 9.3. What is the specific volume of CH_4 at 200 lbf/in and 500°F? Let suffix 1 denote the new condition and suffix 0 denote the standard condition.

Now $\qquad \dfrac{P_1 V_1}{T_1} = \dfrac{P_0 V_0}{T_0}$

Hence
$$
\begin{aligned}
V_1 &= \frac{P_0 T_1 V_0}{P_1 T_0} \\
&= \frac{14\cdot7 \times (460 + 500) \times 22\cdot4}{200 \times (460 + 500)} \\
&= \frac{14\cdot7 \times 960 \times 22\cdot4}{200 \times 492} \\
&= 3\cdot21 \text{ ft}^3/\text{lb}
\end{aligned}
$$

9.5 Theoretical Air

In combustion processes, the oxygen required for the reaction is normally obtained from air. Since air consists mainly of oxygen and nitrogen, then to obtain a certain amount of oxygen for a combustion process requires a much larger amount of air. If the quantity of air supplied during a combustion process provides just sufficient oxygen for all the combustible elements to burn completely, then this quantity of air is known as the theoretical air. The theoretical air to fuel ratio is sometimes termed the stoichiometric ratio.

NOTE

The approximate proportions of oxygen and nitrogen present in the air may be taken as follows:

By weight, air contains 23% oxygen and 77% nitrogen

  i.e. 1 lb of air contains 0·23 lb oxygen and 0·77 lb nitrogen

By volume, air contains 21% oxygen and 79% nitrogen

  i.e. 1 ft³ of air contains 0·21 ft³ oxygen and 0·79 ft³ nitrogen

Although the nitrogen supplied with the oxygen during a combustion process takes no active part in releasing heat energy, it must be included in the analysis of the process.

9.6 Excess Air

Generally speaking, a combustion process has to take place in a limited time. As the full heating value of a fuel can only be obtained if sufficient oxygen is present during the process, it is usual to supply more oxygen than that theoretically required by the process. Normally, this will entail supplying more air than that theoretically required, and the difference between the actual air supplied and the theoretical amount of air required is known as the excess air. The ratio of

$$\frac{\text{actual air supplied} - \text{theoretical air required}}{\text{theoretical air required}} \times 100$$

is known as the percentage excess air.

It has been seen from the combustion equations for hydrogen that the product formed is H_2O. This is normally in the form of vapour (steam), but if the products of combustion are cooled this steam will be condensed to water and may be removed from the products of combustion. In this case the products of combustion are known as the dry products.

9.7 Calorific Value

The calorific value of a fuel is defined as the heat energy which becomes available when unit quantity of the fuel is burned under specified conditions. The two most common methods of determining the calorific values of fuels are described in §§9.8 and 9.9.

9.8 Bomb Calorimeter

If a fuel is in the form of a solid (e.g. coal) or a liquid (e.g. fuel oil) a bomb calorimeter is normally used to determine the calorific value of the fuel. A bomb calorimeter consists of a gas tight vessel of known water equivalent and a larger well-insulated vessel, also of known water equivalent. Fig. 9.1 shows a diagrammatic sketch of a bomb calorimeter. In use, the sample of fuel is carefully weighed in the crucible and the crucible placed in the stirrup inside the bomb, along with a small known quantity of water in the base of the bomb. A piece of ignition wire is then connected between the two electrodes and arranged to pass through the sample of fuel. The cap of the bomb is then carefully screwed into place and the bomb filled with oxygen at about 25 atmospheres from an oxygen bottle before being placed inside the well-insulated vessel. A known quantity of water is now poured into the vessel until the bomb is almost covered, and the stirring attachment and lid fitted. A sensitive thermometer is then inserted through the lid, and conditions are allowed to settle. When the thermometer shows a steady reading, the stirring mechanism is switched on and a stop clock started. Corresponding readings of temperature and time are recorded for about five minutes and at a definite time, say at the end of the fifth

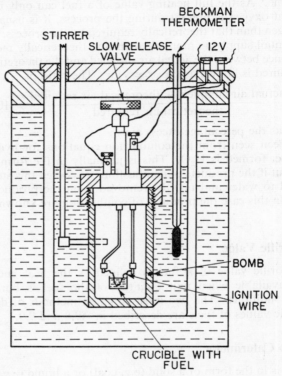

FIGURE 9.1

minute, the electrical firing circuit is completed. Temperatures are then recorded every half minute for about 15 minutes, by which time the temperature of the water should have reached a maximum value and be decreasing again.

The following example will serve to illustrate how the data obtained are used to determine the calorific value of the fuel under test.

Ex. 9.4. During an experiment to determine the calorific value of coal, a bomb calorimeter of water equivalent 465 g was used. From the test observations listed below calculate the calorific value of the coal sample.

Weight of coal used	1·007 g
Weight of water inside the bomb	10·0 g
Weight of water surrounding bomb	2,200 g
Corrected temperature rise after firing	2·95 degC

Total water equivalent of apparatus $= 465 + 10 + 2,200$ g
$$= 2,675 \text{ g}$$
$$= \frac{2,675}{453·6}$$
$$= 5·9 \text{ lb}$$

Corrected temperature rise $= 2·95$ degC
$$= 2·95 \times 1·8$$
$$= 5·3 \text{ degF}$$

Energy absorbed by apparatus $= 5·9$ lb $\times$ 5·3 degF $\times$ 1 Btu/lb degF
$$= 31·3 \text{ Btu}$$

$\therefore$ energy released by 1·007 g coal $= 31·3$ Btu

$\therefore$ energy released per lb of coal $= \dfrac{31·3}{1·007} \times 453·6$
$$= 14,100 \text{ Btu}$$

i.e. calorific value of coal $= 14,100$ Btu/lb

9.9 Gas Calorimeter

If the fuel is in a gaseous form (e.g. town's gas), a gas calorimeter (shown in Fig. 9.2) is used to determine the calorific value. With this type of calorimeter, the fuel passes through a constant pressure regulator and through an accurate gas meter before being burned at constant pressure in the calorimeter burner. The products of combustion are then made to pass over a number of coils through which water is fed from a constant head device, before being allowed to escape into the atmosphere. The water, after passing through the coils, can either be discharged to drain or into a calibrated vessel. Before starting the test, water is poured into the calorimeter until the level of the condensate drain is reached. The water supply to the coils is then turned on, the calorimeter burner lit, and conditions allowed to become steady. When steady conditions have been attained, the following readings are recorded:

(a) Rate of water flow
(b) Inlet and outlet temperatures of the water

(c) Rate of gas consumption
(d) Gas pressure
(e) Exhaust gas temperature
(f) Rate of collection of condensate
(g) Barometric pressure and temperature

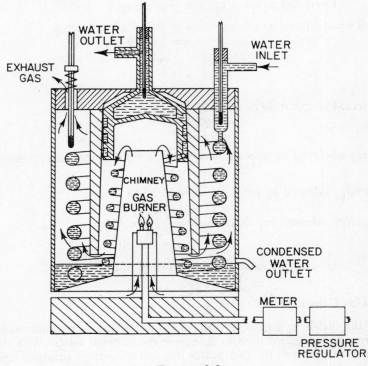

FIGURE 9.2

The following example will serve to illustrate how the calorific value of the fuel can be calculated from the data obtained.

Ex. 9.5. A gas calorimeter was used to determine the calorific value of a certain gas. During the test, the following observations were recorded when conditions were steady:

Rate of gas consumption	4·3 ft³/h
Gas pressure (above atmospheric)	0·6 in water gauge
Gas temperature	63°F
Exhaust gas temperature	64°F
Weight of condensed water collected	26·5 g in 20 min

Rate of water circulation .... 400 g/min
Temperature rise of circulated water ... 19·3 degC
Atmospheric pressure .... 29·17 in mercury
Atmospheric temperature .... 63·5°F

Determine the higher and lower calorific values of the gas.

Absolute pressure of gas = 29·17 inHg + 0·6 in W.G.

$$= \left(29\cdot17 \times \frac{14\cdot7}{30}\right) + \left(\frac{0\cdot6}{12} \times \frac{62\cdot4}{144}\right) \text{lbf/in}^2$$
$$= 14\cdot29 + 0\cdot0217$$
$$= 14\cdot31 \text{ lbf/in}^2$$

∴ at 14·7 lbf/in² and 60°F

Volume of gas burned per hour $= 4\cdot3 \text{ ft}^3 \times \dfrac{14\cdot31 \text{ lbf/in}^2}{14\cdot7 \text{ lbf/in}^2} \times \dfrac{520°\text{R}}{523°\text{R}}$
$$= 4\cdot16 \text{ ft}^3$$

Energy produced per hour $= 400 \text{ g} \times 19\cdot3 \text{ degC} \times 60$
$$= \frac{400}{453\cdot6} \text{ lb} \times (19\cdot3 \times 1\cdot8) \text{ degF} \times 60$$
$$= 1{,}840 \text{ Btu/h}$$

i.e. 4·16 ft³ produce 1,840 Btu

Higher calorific value $= \dfrac{1{,}840}{4\cdot16}$
$$= 442 \text{ Btu/ft}^3$$

H_2O condensed per ft³ of gas burned $= \dfrac{26\cdot5 \times 3}{4\cdot16} \text{ g}$
$$= \frac{26\cdot5 \times 3}{4\cdot16 \times 453\cdot6}$$
$$= 0\cdot0421 \text{ lb}$$

∴ Lower calorific value $= 442 - (0\cdot0421 \times 1{,}050)$ (see §9.10)
$$= 442 - 44\cdot1$$
$$= 397\cdot9 \text{ Btu/ft}^3$$

9.10 Higher and Lower Calorific Values

In both of the foregoing methods the products of combustion are cooled to the original temperature of the fuel, and any steam which has been formed due to the combustion of hydrogen in the fuel will therefore have been condensed, the latent heat given up by this steam being

8

taken up by the cooling water. Under these circumstances the calorific value determined is known as the Higher Calorific Value. In practice, however, it is unusual to cool the products of combustion to the original temperature and hence this latent heat is not normally available. Under these conditions, the calorific value will be lower than that determined by the foregoing methods, and is called the Lower Calorific Value. The expression relating the two calorific values will be

$$(LCV) = (HCV) - m_f \times h_{fg} \qquad (9.13)$$

where m_f = mass of water condensed (lb) and h_{fg} = specific enthalpy of steam above a nominated datum. The British Standard Specification states that in all cases where the higher calorific value is converted to the lower calorific value, the value of h_{fg} is to be taken as 1,050 Btu/lb above a nominated condition of water at 77°F.

Ex. 9.6. A fuel consisting of 84% carbon, 15% hydrogen and 1% oxygen by weight is burned with 30% excess air. Determine the volumetric analysis, and the molecular weight, and characteristic gas constant of the dry exhaust gases.

The Universal Gas Constant is 1·985 Btu/mol degR.
Air contains 23% oxygen and 77% nitrogen by weight.

Consider the combustion of 1 lb fuel, i.e. 0·84 lb carbon + 0·15 lb hydrogen.

Combustion of carbon
Chemical equation $\qquad C + O_2 \rightarrow CO_2$
Mass equation $\qquad$ 12 lb C + 32 lb O_2 → 44 lb CO_2

∴ for 0·84 lb carbon, $\quad$ 0·84 lb C + $\left(\dfrac{32}{12} \times 0·84\right)$ lb O_2

$$\rightarrow \left(\frac{44}{12} \times 0·84\right) \text{lb } CO_2$$

i.e. $\qquad$ 0·84 lb C + 2·24 lb O_2 → 3·08 lb CO_2

Combustion of hydrogen
Chemical equation $\qquad 2H_2 + O_2 \rightarrow 2H_2O$
Mass equation $\qquad$ 4 lb H_2 + 32 lb O_2 → 36 lb H_2O

∴ for 0·15 lb hydrogen, 0·15 lb H_2 + $\left(\dfrac{32}{4} \times 0·15\right)$ lb O_2

$$\rightarrow \left(\frac{36}{4} \times 0·15\right) \text{lb } H_2O$$

i.e. $\qquad$ 0·15 lb H_2 + 1·2 lb O_2 → 1·35 lb H_2O

Therefore the complete combustion of 1 lb fuel requires $(2 \cdot 24 + 1 \cdot 2)$ $= 3 \cdot 44$ lb oxygen. Since 1 lb fuel already contains $0 \cdot 01$ lb oxygen, oxygen required from theoretical air $= 3 \cdot 44 - 0 \cdot 01 = 3 \cdot 43$ lb.

Since air contains 23% oxygen by weight

$$\text{theoretical air} = 3 \cdot 43 \times \frac{100}{23} = 14 \cdot 91 \text{ lb}$$

$\therefore$ $$\text{air supplied} = \text{theoretical air} \times \frac{130}{100}$$

$$= 14 \cdot 91 \times 1 \cdot 3 = 19 \cdot 38 \text{ lb}$$

This will contain $19 \cdot 38 \times \dfrac{23}{100} = 4 \cdot 46$ lb oxygen

and $$19 \cdot 38 \times \frac{77}{100} = 14 \cdot 92 \text{ lb nitrogen}$$

$\therefore$ oxygen remaining unused $= 4 \cdot 46 - 3 \cdot 43 = 1 \cdot 03$ lb

Therefore exhaust gas consists of $3 \cdot 08$ lb CO_2; $1 \cdot 35$ lb H_2O; $1 \cdot 03$ lb O_2; $14 \cdot 92$ lb N_2.

Dry exhaust gas contains $3 \cdot 08$ lb CO_2; $1 \cdot 03$ lb O_2; $14 \cdot 92$ lb N_2.

Since 1 mol of CO_2 weighs 44 lb, $3 \cdot 08$ lb $CO_2 \equiv \dfrac{3 \cdot 08}{44} = 0 \cdot 07$ mol

Since 1 mol of O_2 weighs 32 lb, $1 \cdot 03$ lb $O_2 \equiv \dfrac{1 \cdot 03}{32} = 0 \cdot 0322$ mol

Since 1 mol of N_2 weighs 28 lb, $14 \cdot 92$ lb $N_2 \equiv \dfrac{14 \cdot 92}{28} = 0 \cdot 533$ mol

Therefore dry exhaust gas consists of $0 \cdot 07$ mol $CO_2 + 0 \cdot 0322$ mol O_2 $+ 0 \cdot 533$ mol N_2, i.e. $0 \cdot 6352$ mol total.

Therefore percentage analysis by mols (and therefore by volume) is

$$\frac{0 \cdot 07}{0 \cdot 6352} \times 100\% \text{ } CO_2; \quad \frac{0 \cdot 0322}{0 \cdot 6352} \times 100\% \text{ } O_2; \quad \frac{0 \cdot 533}{0 \cdot 6352} \times 100\% \text{ } N_2$$

i.e. $11 \cdot 14\%$ CO_2; $5 \cdot 06\%$ O_2; $83 \cdot 8\%$ N_2.

Total weight of dry exhaust gas $= 3 \cdot 08 + 1 \cdot 03 + 14 \cdot 92 = 19 \cdot 03$ lb

Total volume of this dry exhaust gas $= 0 \cdot 6352$ mol

Therefore molecular weight of dry exhaust gas (i.e. weight of 1 mol) is

$$\frac{19 \cdot 03}{0 \cdot 6352 \text{ mol}} \text{ lb} = 29 \cdot 9 \text{ lb/mol}$$

Since molecular weight $\times R = $ Universal Gas Constant

$$\therefore \quad 29.9 \, \frac{\text{lb}}{\text{mol}} \times R = 1.985 \, \frac{\text{Btu}}{\text{mol degR}}$$

$$\therefore \quad R = 1.985 \, \frac{\text{Btu}}{\text{mol degR}} \times \frac{1}{29.9} \, \frac{\text{mol}}{\text{lb}}$$

$$= 0.0663 \, \frac{\text{Btu}}{\text{lb degR}}$$

$$= 51.6 \, \frac{\text{ft lbf}}{\text{lb degR}}$$

Ex. 9.7. A fuel gas having a volumetric analysis of 30% CH_4, 10% C_2H_4, 40% H_2, 10% CO, 2% CO_2, 7% N_2, 1% O_2, is burned with a volumetric air:fuel ratio of 6:1. Determine the analysis of the dry exhaust gases by weight and by volume.

Air contains 21% oxygen and 79% nitrogen by volume.

Consider 1 lb mol of fuel. This will contain

$$\frac{30}{100} = 0.3 \text{ mol } CH_4; \quad \frac{10}{100} = 0.1 \text{ mol } C_2H_4; \quad \frac{40}{100} = 0.4 \text{ mol } H_2;$$

$$\frac{10}{100} = 0.1 \text{ mol } CO; \quad \frac{2}{100} = 0.02 \text{ mol } CO_2; \quad \frac{7}{100} = 0.07 \text{ mol } N_2;$$

$$\frac{1}{100} = 0.01 \text{ mol } O_2.$$

CH_4, C_2H_4, H_2, and CO are combustibles.

O_2 will be utilised in the combustion process. CO_2 and N_2 will not take any active part in the combustion process and will pass through to the exhaust.

Combustion of CH_4

Chemical equation	$CH_4 + 2O_2 \rightarrow CO_2 + 2H_2O$
'Mol' equation	1 mol CH_4 + 2 mol O_2
	$\rightarrow$ 1 mol CO_2 + 2 mol H_2O

Since the fuel contains 0.3 mol CH_4, multiplying through by 0.3 gives

$$0.3 \text{ mol } CH_4 + 0.6 \text{ mol } O_2$$
$$\rightarrow 0.3 \text{ mol } CO_2 + 0.6 \text{ mol } H_2O$$

Combustion of C_2H_4

Chemical equation	$C_2H_4 + 3O_2 \rightarrow 2CO_2 + 2H_2O$
'Mol' equation	1 mol C_2H_4 + 3 mol O_2
	$\rightarrow$ 2 mol CO_2 + 2 mol H_2O

Since the fuel contains 0.1 mol C_2H_4, multiplying through by 0.1 gives

$$0.1 \text{ mol } C_2H_4 + 0.3 \text{ mol } O_2$$
$$\rightarrow 0.2 \text{ mol } CO_2 + 0.2 \text{ mol } H_2O$$

Combustion of H_2
Chemical equation $2H_2 + O_2 \rightarrow 2H_2O$
'Mol' equation $2 \text{ mol } H_2 + 1 \text{ mol } O_2 \rightarrow 2 \text{ mol } H_2O$

Since the fuel contains $0.4 \text{ mol } H_2$,

$$0.4 \text{ mol } H_2 + 0.2 \text{ mol } O_2 \rightarrow 0.4 \text{ mol } H_2O$$

Combustion of CO
Chemical equation $2CO + O_2 \rightarrow 2CO_2$
'Mol' equation $2 \text{ mol } CO + 1 \text{ mol } O_2 \rightarrow 2 \text{ mol } CO_2$

Since the fuel contains $0.1 \text{ mol } CO$,

$$0.1 \text{ mol } CO + 0.05 \text{ mol } O_2 \rightarrow 0.1 \text{ mol } CO_2$$

Therefore oxygen required for combustion of 1 mol of fuel $=$ $0.6 + 0.3 + 0.2 + 0.05 = 1.15$ mol. The fuel already contains 0.01 mol of oxygen, and therefore oxygen required from the air supplied $=$ $1.15 - 0.01 = 1.14$ mol. Since air consists of $21\% \text{ } O_2$ and $79\% \text{ } N_2$ by volume (i.e. by mol) the 6 mol of air supplied per mol of fuel will contain $6 \times 0.21 = 1.26 \text{ mol } O_2$ and $6 \times 0.79 = 4.74 \text{ mol } N_2$.

If $1.14 \text{ mol } O_2$ are required for combustion, then $1.26 - 1.14$ $= 0.12 \text{ mol } O_2$ will remain unused and appear in the exhaust gas.

The N_2 in the air supplied must also appear in the exhaust gas. Therefore the exhaust gas will consist of

$$(0.3 + 0.2 + 0.1 + 0.02) \text{ mol } CO_2; \quad (0.6 + 0.2 + 0.4) \text{ mol } H_2O;$$
$$(4.74 + 0.07) \text{ mol } N_2; \quad 0.12 \text{ mol } O_2.$$

i.e. $0.62 \text{ mol } CO_2 + 1.2 \text{ mol } H_2O + 4.81 \text{ mol } N_2 + 0.12 \text{ mol } O_2$.

The *dry* exhaust gas will consist of

$$0.62 \text{ mol } CO_2 + 4.81 \text{ mol } N_2 + 0.12 \text{ mol } O_2$$
i.e. 5.55 mol total

Therefore percentage analysis by volume (or mol) will be

$$\frac{0.62}{5.55} \times 100\% \text{ } CO_2; \quad \frac{4.81}{5.55} \times 100\% \text{ } N_2; \quad \frac{0.12}{5.55} \times 100\% \text{ } O_2$$

i.e. $11.18\% \text{ } CO_2; \quad 86.6\% \text{ } N_2; \quad 2.16\% \text{ } O_2$

i.e. 1 lb mol of dry exhaust gas contains $0.1118 \text{ mol } CO_2$, $0.866 \text{ mol } N_2$ and $0.0216 \text{ mol } O_2$.

Since 1 mol of CO_2 weighs 44 lb (molecular weight of $CO_2 = 44$)

$$0.1118 \text{ mol } CO_2 \text{ weigh } 0.1118 \times 44 = 4.92 \text{ lb}$$

similarly $0.866 \text{ mol } N_2 \text{ weigh } 0.866 \times 28 = 24.25 \text{ lb}$

and $0.0216 \text{ mol } O_2 \text{ weigh } 0.0216 \times 32 = 0.691 \text{ lb}$

Therefore total weight of 1 mol of dry exhaust gas (i.e. its molecular weight) is $4 \cdot 92 + 24 \cdot 25 + 0 \cdot 691 = 29 \cdot 861$ lb and percentage analysis by weight will be given by

$$\frac{4 \cdot 92}{29 \cdot 861} \times 100 \% \text{ CO}_2; \quad \frac{24 \cdot 25}{29 \cdot 861} \times 100 \% \text{ N}_2; \quad \frac{0 \cdot 691}{29 \cdot 861} \times 100 \% \text{ O}_2$$

i.e. $16 \cdot 45 \%$ CO_2; $81 \cdot 2 \%$ N_2; $2 \cdot 315 \%$ O_2.

Ex. 9.8. A sample of coal having a calorific value of 12,800 Btu/lb has a composition by weight of:

$$\text{C } 80 \%; \text{ H}_2 \text{ } 4 \%; \text{ O}_2 \text{ } 5 \%; \text{ S } 2 \%; \text{ N}_2 \text{ } 3 \%; \text{ remainder ash}$$

When the coal was burned in a furnace with 60% excess air the temperature of the flue gases entering the stack was $600 °F$. For an atmospheric temperature of $60 °F$, determine

(a) The actual weight of air supplied per lb of fuel
(b) The percentage of the heat energy released that is carried away by the flue gases

Assume combustion to be complete, and that the heat energy carried away per lb of moisture is 1,300 Btu. For the dry flue gases, take $c_p = 0 \cdot 25$ Btu/lb degR. Air contains 23% O_2 and 77% N_2 by weight.

Consider 1 lb of fuel
For complete combustion of carbon

$$C + O_2 \rightarrow CO_2$$

i.e. $\qquad$ 12 lb C $+$ 32 lb $O_2 \rightarrow$ 44 lb CO_2

∴ $\qquad$ 1 lb C $+$ 2·67 lb $O_2 \rightarrow$ 3·67 lb CO_2

∴ $\qquad$ 0·8 lb C $+$ 2·14 lb $O_2 \rightarrow$ 2·94 lb CO_2

For complete combustion of hydrogen

$$2H_2 + O_2 \rightarrow 2H_2O$$

i.e. $\qquad$ 4 lb H_2 $+$ 32 lb $O_2 \rightarrow$ 36 lb H_2O

∴ $\qquad$ 1 lb H_2 $+$ 8 lb $O_2 \rightarrow$ 9 lb H_2O

∴ $\qquad$ 0·04 lb H_2 $+$ 0·32 lb $O_2 \rightarrow$ 0·36 lb H_2O

For complete combustion of sulphur

$$S + O_2 \rightarrow SO_2$$

i.e. $$32 \text{ lb S} + 32 \text{ lb O}_2 \rightarrow 64 \text{ lb SO}_2$$

$\therefore$ $$1 \text{ lb S} + 1 \text{ lb O}_2 \rightarrow 2 \text{ lb SO}_2$$

$\therefore$ $$0\cdot02 \text{ lb S} + 0\cdot02 \text{ lb O}_2 \rightarrow 0\cdot04 \text{ lb SO}_2$$

$\therefore$ theoretical oxygen required per lb of fuel $= 2\cdot14 + 0\cdot32 + 0\cdot02$
$$= 2\cdot48 \text{ lb}$$

But 1 lb fuel already contains $0\cdot05$ lb oxygen.

$\therefore$ theoretical oxygen required from air $= 2\cdot48 - 0\cdot05$
$$= 2\cdot43 \text{ lb}$$

$\therefore$ theoretical air required $= 2\cdot43 \times \dfrac{100}{23}$
$$= 10\cdot58 \text{ lb}$$

$\therefore$ actual air supplied $= 10\cdot58 \times \dfrac{160}{100}$
$$= 16\cdot9 \text{ lb}$$

This will consist of

$$16\cdot9 \times \frac{23}{100} = 3\cdot89 \text{ lb oxygen}$$

and $$16\cdot9 \times \frac{77}{100} = 13\cdot01 \text{ lb nitrogen}$$

Of this $3\cdot89$ lb of oxygen, $2\cdot43$ lb are used during combustion.

$\therefore$ weight of oxygen in flue gases $= 3\cdot89 - 2\cdot43$
$$= 1\cdot46 \text{ lb}$$

weight of nitrogen in flue gases $=$ weight of nitrogen in fuel
$$+ \text{ weight of nitrogen supplied}$$
$$= 0\cdot03 \text{ lb} + 13\cdot01 \text{ lb}$$
$$= 13\cdot04 \text{ lb}$$

$\therefore$ flue gases consist of

(a) Dry products: $1\cdot46$ lb oxygen $+ 13\cdot04$ lb nitrogen $+ 2\cdot94$ lb CO_2 $+ 0\cdot04$ lb $SO_2 = 17\cdot48$ lb

(b) Moisture $= 0\cdot36$ lb H_2O

Heat energy carried away by

(a) Dry products $= 17\cdot48 \text{ lb} \times 0\cdot25 \dfrac{\text{Btu}}{\text{lb degR}} \times (600 - 60) \text{ degR}$
$$= 2,360 \text{ Btu}$$

(b) Moisture $= 0.36 \text{ lb} \times 1,300 \dfrac{\text{Btu}}{\text{lb}} = 468 \text{ Btu}$

$\therefore$ total heat energy carried away per lb of fuel $= 2,360 + 468$
$$= 2,828 \text{ Btu}$$

Heat energy released per lb of fuel $= 12,800 \text{ Btu}$

$\therefore$ percentage loss $= \dfrac{2,828}{12,800} = 22.1\%$

Ex. 9.9. A gaseous fuel has a volumetric composition of:

$$CH_4 \; 10\%; \; H_2 \; 24\%; \; O_2 \; 3\%; \; CO \; 12\%; \; N_2 \; 51\%$$

The gas is burned with a volumetric air:fuel ratio of $5:1$. Determine

(a) The percentage analysis by volume of the combustion products
(b) The mean molecular weight of the dry products
(c) The characteristic gas constant of the dry products

Air contains 21% oxygen and 79% nitrogen by volume. $R_0 = 1,545$ ft lbf/mol degR.

(a) *Consider* 1 mol *of fuel*
     For combustion of CH_4

$$CH_4 + 2O_2 \rightarrow CO_2 + 2H_2O$$

i.e. $1 \text{ mol } CH_4 + 2 \text{ mol } O_2 \rightarrow 1 \text{ mol } CO_2 + 2 \text{ mol } H_2O$

$\therefore$ $0.1 \text{ mol } CH_4 + 0.2 \text{ mol } O_2 \rightarrow 0.1 \text{ mol } CO_2 + 0.2 \text{ mol } H_2O$

For combustion of H_2

$$2H_2 + O_2 \rightarrow 2H_2O$$

i.e. $2 \text{ mol } H_2 + 1 \text{ mol } O_2 \rightarrow 2 \text{ mol } H_2O$

$\therefore$ $1 \text{ mol } H_2 + 0.5 \text{ mol } O_2 \rightarrow 1 \text{ mol } H_2O$

$\therefore$ $0.24 \text{ mol } H_2 + 0.12 \text{ mol } O_2 \rightarrow 0.24 \text{ mol } H_2O$

For combustion of CO

$$2CO + O_2 \rightarrow 2CO_2$$

i.e. $2 \text{ mol } CO + 1 \text{ mol } O_2 \rightarrow 2 \text{ mol } CO_2$

$\therefore$ $1 \text{ mol } CO + 0.5 \text{ mol } O_2 \rightarrow 1 \text{ mol } CO_2$

$\therefore$ $0.12 \text{ mol } CO + 0.06 \text{ mol } O_2 \rightarrow 0.12 \text{ mol } CO_2$

$\therefore$ oxygen required per mol of fuel $= 0.2 + 0.12 + 0.06 = 0.38 \text{ mol}$

Since 1 mol of fuel already contains 0·03 mol O_2 then oxygen required from air = 0·38 − 0·03 = 0·35 mol.

But 5 mol of air are supplied per mol of fuel, i.e. 5 × 0·21 = 1·05 mol oxygen and 5 × 0·79 = 3·95 mol nitrogen.

Therefore oxygen remaining unused = 1·05 − 0·35 = 0·7 mol.

Therefore exhaust products consist of:

	CO_2	H_2O	O_2	N_2	Total
Mol	(0·1 + 0·12) = 0·22	(0·2 + 0·24) = 0·44	0·7	(0·51 + 3·95) = 4·46	5·82
% vol	$\frac{0·22}{5·82} \times 100$ = 3·78%	$\frac{0·44}{5·82} \times 100$ = 7·56%	$\frac{0·7}{5·82} \times 100$ = 12·02%	$\frac{4·46}{5·82} \times 100$ = 76·5%	99·86%

(b) *To calculate the mean molecular weight*

Dry exhaust gas consists of

	CO_2	O_2	N_2	Total
Mol	0·22	0·7	4·46	5·38
Weight of 1 mol (i.e. molecular wt)	44	32	28	
Weight in exhaust gas	0·22 × 44 = 9·68 lb	0·7 × 32 = 22·4 lb	4·46 × 28 = 125 lb	157·08 lb

i.e. 5·38 mol weigh 157·08 lb

∴ molecular weight = $\frac{157·08}{5·38}$ = 29·2

(c) Characteristic Gas Constant = $\dfrac{\text{Universal Gas Constant}}{\text{molecular weight}}$

$= \dfrac{1,545 \text{ ft lbf/mol degR}}{29·2 \text{ lb/mol}}$

= 52·9 ft lbf/lb degR

EXERCISES ON CHAPTER 9

Air contains 23% O_2, 77% N_2 by weight, or 21% O_2, 79% N_2 by volume.

1. A boiler fuel has an analysis by weight of 77% carbon, 6% hydrogen, 6% oxygen, 2% sulphur, the remainder being incombustible. When the fuel is burned, 60% excess air is supplied. Determine the weight of air supplied per lb of fuel, and the weight analysis of the dry flue gases.

(17·35 lb; 15·93% CO_2, 0·23% SO_2, 8·45% O_2, 75·35% N_2)

2. A gaseous fuel has the following analysis by volume: 5% CO, 55% H_2, 25% CH_4, 2% O_2, 3% CO_2, 10% N_2. During combustion 50% excess air is supplied. Determine the volume of air supplied per ft³ of fuel gas, and the volumetric analysis of the exhaust gases.

(5·5 ft³; 5·25% CO_2, 16·7% H_2O, 6·2% O_2, 71·8% N_2)

3. The composition of a gas by volume is 11% CO, 45% H_2, 35% CH_4, 4% CO_2, 3% O_2, and 2% N_2. The gas is mixed with 100% excess air for use in a gas engine. Determine the volume of air supplied per ft³ of gas and the percentage analysis by volume of the dry exhaust gas.

(9·05 ft³; 5·8% CO_2, 11·02% O_2, 83·2% N_2)

4. The coal used by a boiler has a percentage analysis by weight of 88% carbon, 6% hydrogen, 4% oxygen, the remainder being incombustible. Calculate the stoichiometric air:fuel ratio. If 18 lb of air are actually supplied per lb of fuel, determine the percentage excess air, and the analysis by weight of the wet exhaust products.

(12·13:1; 48·4%; 17% CO_2, 2·85% H_2O, 7·1% O_2, 73% N_2)

5. An internal combustion engine operating on heptane as fuel has an air:fuel ratio of 23:1 by weight. Determine the percentage analysis by volume of the dry products of combustion. (Heptane $\equiv C_7H_{16}$).

(9·25% CO_2, 7·3% O_2, 83·4% N_2)

6. A volumetric analysis of a fuel gas showed the following composition: 50% H_2, 8% CO, 2% CO_2, 2% O_2, 5% N_2, 28% CH_4, 5% C_4H_8. If the fuel is burned with an air:fuel ratio of 6:1, determine the percentage analysis of the wet products by volume and by weight.

(By volume: 8·51% CO_2, 18·7% H_2O, 1·93% O_2, 71·1% N_2)
(By weight: 13·48% CO_2, 12·22% H_2O, 2·25% O_2, 72·3% N_2)

7. Fuel having an analysis of 82% carbon, 12% hydrogen, and 6% ash by weight, is supplied to a furnace at the rate of 1·8 ton/h. The air is supplied to the furnace at 15 lbf/in² and 80°F at the rate of 300 ft³/sec. If the characteristic gas constant for air is 53 ft lbf/lb deg R, calculate the percentage excess air supplied, and the volumetric analysis of the dry exhaust gases, assuming combustion to be complete.

(47·0%; 10·2% CO_2, 7% O_2, 82·8% N_2)

8. A boiler uses coal having a weight analysis of 82% carbon, 4% hydrogen, 8% oxygen, 0·5% sulphur, and the remainder ash. If 20 lb of air are supplied per lb of coal, determine the analysis of the dry flue gases by weight and by volume.

(By weight: 14·6% CO_2, 0·48% SO_2, 10·4% O_2, 74·8% N_2)
(By volume: 9·94% CO_2, 0·02% SO_2, 9·89% O_2, 80·1% N_2)

10: Reciprocating Gas Compressors

Gas compression is an important process which is used in many branches of engineering. The term is generally applied to processes involving considerable increases of gas pressure and gas density. In the majority of processes the gas used is air, but the term gas compression is used in preference to air compression.

The industrial applications are manifold and occur in connection with such processes as compressed air motors for driving pneumatic tools, automatic control devices, air brakes, material conveyance along pipes, bottled gas production, chemical engineering processes and many other applications.

10.1 Types of Compressors

Gas compressors may be sub-divided into two distinct types:
(a) The positive displacement type such as the reciprocating gas compressor and the rotary positive displacement compressor, e.g. Roots blower, vane-sealed compressors and gear pumps
(b) The rotary compressors of the turbine type, such as axial flow compressors and centrifugal compressors

In the following sections, the reciprocating compressor only will be discussed.

10.2 The Ideal Reciprocating Compressor

Fig. 10.1 shows a $P-V$ diagram for an ideal reciprocating gas compressor having no clearance volume. The cycle consists of the following processes:

A–B   The suction stroke, where the piston moving down the cylinder causes the suction valve to open and a volume of gas V_1 ft^3 at a pressure of P_1 lbf/ft^2 is drawn into the cylinder. Since the gas is flowing across the boundary into the system, the gas will do work on the piston equal to the area under AB during this stroke
B–C   Immediately the piston starts to return along the cylinder, the pressure of the gas inside the cylinder begins to rise, causing the

suction valve to close and the gas to be sealed within the cylinder. Further movement of the piston raises the pressure of the gas until the delivery pressure P_2 lbf/ft² is reached at C where the pressure of the gas is sufficient to force open the delivery valve. If the process BC is assumed to be polytropic, following the

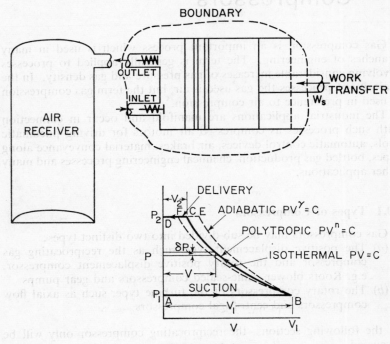

FIGURE 10.1

law $PV^n =$ constant, then the work done on the gas by the piston during this portion of the return stroke will be equal to the area under BC

C–D   With the delivery valve open, further movement of the piston along the cylinder forces the gas out of the cylinder into a receiver at a constant pressure P_2 lbf/ft². The work done on the gas by the piston during this portion of the stroke will be equal to the area under CD

If W_n is the net work energy transferred during this cycle then

$$W_n = \text{area under DC} + \text{area under CB} - \text{area under BA}$$
$$= \text{ABCD}$$

i.e. $\quad W_n = \int_{V_1}^{V_2} V\,dP$

$$= \frac{n}{n-1}(P_1 V_1 - P_2 V_2)\ \text{ft lbf} \tag{10.1}$$

(see Appendix 2.3).

This expression will give a negative value as would be expected since work energy has to flow into the system to compress the gas. An alternative expression for W_n may be obtained as follows:

$$W_n = \frac{n}{n-1}(P_1 V_1 - P_2 V_2)$$

$$= \frac{n}{n-1} P_1 V_1 \left(1 - \frac{P_2 V_2}{P_1 V_1}\right)$$

Since $P_1 V_1{}^n = P_2 V_2{}^n$ then

$$\frac{V_2}{V_1} = \left(\frac{P_1}{P_2}\right)^{1/n} = \left(\frac{P_2}{P_1}\right)^{-1/n}$$

$\therefore \qquad W_n = \frac{n}{n-1} P_1 V_1 \left[1 - \left(\frac{P_2}{P_1}\right)\left(\frac{P_2}{P_1}\right)^{-1/n}\right]$

$$= \frac{n}{n-1} P_1 V_1 \left[1 - \left(\frac{P_2}{P_1}\right)^{(n-1)/n}\right] \tag{10.2}$$

This is a convenient form since initial conditions P_1 and V_1, and the final pressure P_2 are generally known. If the characteristic gas equation is applied to equation (10.1) an expression for W_n will be obtained in terms of the initial and final temperatures as follows:

$$W_n = \frac{n}{n-1}(P_1 V_1 - P_2 V_2)$$

Since $P_1 V_1 = mRT_1$ and $P_2 V_2 = mRT_2$ then

$$W_n = \frac{n}{n-1}(mRT_1 - mRT_2)$$

$$= \frac{n}{n-1} mR(T_1 - T_2)$$

$$= \frac{n}{n-1} mRT_1 \left(1 - \frac{T_2}{T_1}\right) \tag{10.3}$$

10.3 Isentropic Compression

In reciprocating air compressors the value of n is usually about 1·3. If the compression process is isentropic (i.e. reversible adiabatic) it will follow the equation $PV^{\gamma} = $ constant, where $\gamma = c_p/c_v = 1.4$ for air, and be represented by the line BE in Fig. 10.1. The fact that this isentropic curve is steeper than the usual polytropic curve may be verified by differentiating the equation of the curve $PV^n = $ constant as follows:

$$PV^n = \text{constant}$$

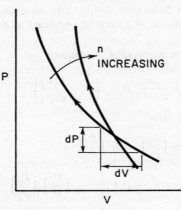

FIGURE 10.2

Differentiating with respect to V gives

$$\frac{dP}{dV} V^n + PnV^{n-1} = 0$$

$$\therefore \qquad\qquad \frac{dP}{dV} = - n\frac{P}{V} \qquad\qquad (10.4)$$

i.e. the slope of the curve is negative and becomes steeper with increasing values of n (see Fig. 10.2). For isentropic compression the expression for the work energy required per cycle W_{γ} is given by substituting γ for n in equation (10.2), i.e.

$$W_{\gamma} = \frac{\gamma}{\gamma - 1} P_1 V_1 \left\{ 1 - \left(\frac{P_2}{P_1}\right)^{(\gamma-1)/\gamma} \right\} \qquad (10.5)$$

10.4 Isothermal Compression

If the compression process is isothermal, i.e. it follows the equation $PV = $ constant, it can be seen from equation (10.4) that the compression curve is less steep than the polytropic curve and will therefore lie below it as shown by the line BF in Fig. 10.1. In this case the expression for the work energy required per cycle (W_i) is given by:

$$W_i = P_1 V_1 \ln \frac{P_1}{P_2} \qquad (10.6)$$

(see Appendix 2.3.)

It can be seen from the P–V diagrams in Fig. 10.1 that the work energy required per cycle by a compressor having an isothermal compression process is less than that required if the compression process is either reversible polytropic or isentropic. As a result of the presence of friction in an actual compressor, the work energy required per cycle will be larger than that required in an ideal compressor. In practice the area of the P–V diagram gives a close approximation of the work energy required, and it is implicit that a certain amount of cooling of the cylinder walls is required if the index of compression n is to be less than the isentropic index.

10.5 Efficiencies of a Reciprocating Compressor

For a compressor with reversible polytropic compression, the work energy required per cycle was given by equation (10.2) as

$$W_n = \frac{n}{n-1} P_1 V_1 \left\{ 1 - \left(\frac{P_2}{P_1} \right)^{(n-1)/n} \right\}$$

where V_1 = volume of gas drawn into the cylinder per cycle. If N = number of completed cycles per minute then the volume of air drawn into the cylinder per minute (V_1') will be given by:

$$V_1' = N V_1$$

and the work energy required per minute by:

$$W_m = \frac{n}{n-1} P_1 V_1 \left\{ 1 - \left(\frac{P_2}{P_1} \right)^{(n-1)/n} \right\} \times N$$

$$= \frac{n}{n-1} P_1 V_1' \left\{ 1 - \left(\frac{P_2}{P_1} \right)^{(n-1)/n} \right\} \text{ ft lbf/min} \qquad (10.7)$$

$\therefore$ polytropic horsepower required $= \dfrac{W_n}{33,000}$

$$= \frac{n}{n-1} \frac{P_1 V_1'}{33,000} \left\{ 1 - \left(\frac{P_2}{P_1} \right)^{(n-1)/n} \right\} \qquad (10.8)$$

In an air compressor the name given to this quantity is air horsepower (ahp).

Similarly for a compressor with isothermal compression, the horse-power required would be given by:

$$\text{Isothermal horsepower} = P_1 V_1 \ln \frac{P_1}{P_2} \times \frac{N}{33,000}$$

$$= \frac{P_1 V_1'}{33,000} \ln \frac{P_1}{P_2} \qquad (10.9)$$

This isothermal horsepower is used as a standard for comparing reciprocating compressors. In this connection, the isothermal efficiency of a compressor is given by:

$$\text{Isothermal efficiency} = \eta_i$$

$$= \frac{\text{isothermal horsepower}}{\text{polytropic horsepower}} \qquad (10.10)$$

In an actual compressor, owing to the effect of bearing friction, windage, etc., the actual horsepower required to drive the compressor will be greater than the polytropic horsepower. This is taken into account by use of a mechanical efficiency for the compressor, defined as:

$$\text{Mechanical efficiency} = \eta_m$$

$$= \frac{\text{polytropic horsepower}}{\text{horsepower input to compressor}} \qquad (10.11)$$

The overall isothermal efficiency of a compressor is defined as:

$$\text{Overall isothermal efficiency} = \eta_0$$

$$= \frac{\text{isothermal horsepower}}{\text{horsepower input to compressor}}$$

$$= \frac{\text{isothermal horsepower}}{\text{polytropic horsepower}} \times \frac{\text{polytropic horsepower}}{\text{horsepower input to compressor}}$$

$$= \eta_i \times \eta_m \qquad (10.12)$$

Ex. 10.1. An air compressor has an effective swept volume of 180 ft³/min and delivers to a receiver at a pressure of 90 lbf/in². The equation of compression is $PV^{1\cdot25} = C$. The temperature of the air at the end of the suction stroke is 90°F and the pressure is 15 lbf/in². Calculate:

(a) The weight of air compressed per min
(b) The temperature at the end of compression
(c) The air horsepower
(d) The isothermal efficiency

Take R for air as $53 \cdot 3$ ft lbf/lb degR.

(a) Weight of air $= \dfrac{P_1 V_1}{R T_1} = \dfrac{144 \times 15 \times 180}{53 \cdot 3 \times (460 + 90)} = 13 \cdot 28$ lbf/min

(b) $\dfrac{T_2}{T_1} = \left(\dfrac{P_2}{P_1}\right)^{(n-1)/n} = \left(\dfrac{90}{15}\right)^{0 \cdot 25/1 \cdot 25} = 6^{0 \cdot 2} = 1 \cdot 431$

$T_2 = (460 + 90)\, 1 \cdot 431 = 550 \times 1 \cdot 431 = 787°\text{R}$

or temperature at end of compression $= 327°\text{F}$

(c) Work done per min $= \dfrac{n}{n-1} P_1 V_1 \left\{ \left(\dfrac{P_2}{P_1}\right)^{(n-1)/n} - 1 \right\}$

$\qquad = \dfrac{1 \cdot 25}{0 \cdot 25} \times 144 \times 15 \times 180 \{1 \cdot 431 - 1\}$

$\qquad = 5 \times 144 \times 15 \times 180 \times 0 \cdot 431$

$\qquad = 837{,}000$ ft lbf

$\qquad$ air horsepower $= \dfrac{837{,}000}{33{,}000}$

$\qquad = 25 \cdot 4$

(d) Isothermal work done per min $= P_1 V_1 \ln \dfrac{P_2}{P_1}$

$\qquad = 144 \times 15 \times 180 \times \ln 6$

$\qquad = 144 \times 15 \times 180 \times 1 \cdot 79$

$\qquad = 696{,}000$ ft lbf

$\qquad$ isothermal horsepower $= \dfrac{696{,}000}{33{,}000}$

$\qquad = 21 \cdot 1$

$\qquad$ isothermal efficiency $= \dfrac{21 \cdot 1}{25 \cdot 4}$

$\qquad = 0 \cdot 83$ or 83%

10.6 Rating of a Compressor

Compressors are often rated by the amount of free air aspirated, compressed and delivered per minute. Free air is air at the pressure and temperature of the surrounding atmosphere near the intake of the compressor and for various reasons (see §10.7) the volume of free air delivered will be less than the swept volume in an actual compressor.

10.7 Effect of Clearance

In practice, all reciprocating compressors must have a clearance volume between the cylinder head and the piston at the end of the delivery stroke. The effect of this clearance volume on the P–V diagram is shown in Fig. 10.3. As the delivery stroke is completed and the piston starts to return along the cylinder, the gas trapped in the clearance volume V_c will expand and undergo a drop in pressure. Until the pressure of this gas reaches the suction pressure (i.e. at 4), no gas can be aspirated through the suction valve. A portion of the swept volume

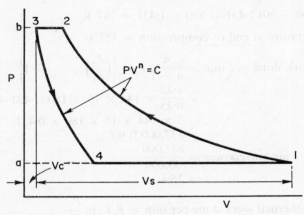

FIGURE 10.3

is therefore, not used to aspirate a further supply of gas into the cylinder. The ratio $(V_1 - V_4)/V_s$ is sometimes called the diagram volumetric efficiency η_{vd}, i.e.

$$\eta_{vd} = \frac{V_1 - V_4}{V_s} = \frac{(V_c + V_s) - V_4}{V_s} \quad (10.13)$$

Since $P_1 V_4{}^n = P_2 V_c{}^n$ then $V_4 = V_c \left(\frac{P_2}{P_1}\right)^{1/n}$

$$\therefore \quad \eta_{vd} = \frac{(V_c + V_s) - V_4}{V_s}$$

$$= \frac{V_c + V_s - V_c \left(\frac{P_2}{P_1}\right)^{1/n}}{V_s}$$

$$= \frac{V_s + V_c \left[1 - \left(\frac{P_2}{P_1}\right)^{1/n}\right]}{V_s}$$

$$= 1 + \frac{V_c}{V_s}\left[1 - \left(\frac{P_2}{P_1}\right)^{1/n}\right] \quad (10.14)$$

The ratio V_c/V_s is known as the clearance ratio, and is normally constant for a particular compressor as it depends on the construction of the compressor. It will be seen from the above expression therefore,

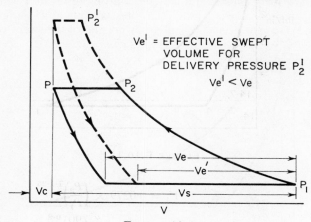

FIGURE 10.4

that with a fixed suction pressure P_1 in a compressor an increase in delivery pressure P_2 will result in a decrease in η_{vd} with less gas being aspirated and delivered per stroke. This is also shown in the P–V diagram in Fig. 10.4.

Ex. 10.2. A single-acting, single-cylinder compressor has cylinder dimensions of 8 in bore and 12 in stroke. The speed is 300 rev/min. The pressure and temperature of the air at the end of the suction stroke is 14 lbf/in² and 85°F and the pressure after compression is 90 lbf/in². The equation of compression and expansion may be taken as $PV^{1\cdot25} = C$. The clearance volume is 8% of the stroke volume. Determine

   (i) The diagram volumetric efficiency
   (ii) The air horsepower
   (iii) The volume of free air aspirated per min if the atmospheric pressure and temperature are 14·7 lb/in² and 55°F

$$R \text{ for air} = 53\cdot3 \text{ ft lbf/lb degR}$$

$$\text{stroke volume} = \frac{\pi}{4} \times \frac{8^2 \times 12}{1,728}$$

$$= 0\cdot349 \text{ ft}^3$$

$$\text{clearance volume} = \frac{8}{100} \times 0\cdot349$$

$$= 0\cdot0279 \text{ ft}^3$$

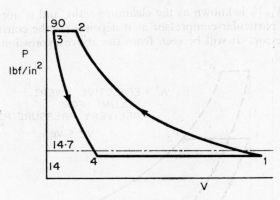

FIGURE E.10.1

$$V_4 = 0.0279 \times \left(\frac{P_2}{P_1}\right)^{1/1.25}$$

$$= 0.0279 \times \left(\frac{90}{14}\right)^{0.8}$$

$$= 0.0279 \times 4.44$$

$$= 0.1238 \ \text{ft}^3$$

$$\text{diagram volumetric efficiency} = \frac{V_1 - V_4}{V_1 - V_3}$$

$$= \frac{(0.349 + 0.0279) - 0.1238}{0.349}$$

$$= \frac{0.2531}{0.349}$$

$$= 0.725$$

$$\text{Mass of air aspirated per stroke} = \frac{P_1(V_1 - V_4)}{RT_1}$$

$$= \frac{144 \times 14 \times 0.2531}{53.3 \times (460 + 85)}$$

$$= \frac{144 \times 14 \times 0.2531}{53.3 \times 545}$$

$$= 0.0176 \ \text{lb}$$

$$\text{Mass of air aspirated per min} = 0.0176 \times 300$$

$$= 5.28 \ \text{lb}$$

$$\text{work done per min} = \frac{n}{n-1} mR(T_1 - T_2)$$

Now,
$$\frac{T_2}{T_1} = \left(\frac{P_2}{P_1}\right)^{(n-1/n)}$$

∴
$$T_2 = 545 \times \left(\frac{90}{14}\right)^{0\cdot25/1\cdot25}$$
$$= 545 \times 1\cdot451$$
$$= 791°R$$

work done per min $= \dfrac{1\cdot25}{0\cdot25} \times 5\cdot28 \times 53\cdot3\,(545 - 791)$

$$= 5 \times 5\cdot28 \times 53\cdot3 \times (-246)$$
$$= -345,000 \text{ ft lbf}$$

air h.p. $= \dfrac{345,000}{33,000}$

$$= 10\cdot46$$

If $V_0 =$ volume of free air

$$\frac{P_0 V_0}{T_0} = \frac{P_1 V_1}{T_1}$$

or
$$V_0 = \frac{P_1 V_1}{T_1} \times \frac{T_0}{P_0}$$
$$= \frac{14}{545} \times \frac{515}{14\cdot7} \times 0\cdot2531$$
$$= 0\cdot2275 \text{ ft}^3/\text{stroke}$$

or $0\cdot2275 \times 300 = 68\cdot25 \text{ ft}^3/\text{min}.$

10.8 Work Energy Required for a Compressor with Clearance

With reference to Fig. 10.3

work energy required per cycle = area 1 2 3 4
  = area 1 2 ba − area 4 3 ba

Assuming that compression and re-expansion processes have the same polytropic index n, then

$$\text{area 1 2 ba} = \frac{n}{n-1}\,P_1 V_1\left[1 - \left(\frac{P_2}{P_1}\right)^{(n-1)/n}\right]$$

and area 4 3 ba $= \dfrac{n}{n-1}\,P_4 V_4\left[1 - \left(\dfrac{P_2}{P_1}\right)^{(n-1)/n}\right]$

also, $P_4 = P_1$

∴ work energy required per cycle = area 1 2 ba − area 4 3 ba

$$= \frac{n}{n-1}\,P_1(V_1 - V_4)\left\{1 - \left(\frac{P_2}{P_1}\right)^{(n-1)/n}\right\} \qquad (10.15)$$

But $(V_1 - V_4)$ equals the volume of gas aspirated at suction conditions during the cycle = effective swept volume.

Applying the characteristic gas equation gives

$$P_1(V_1 - V_4) = mRT_1$$

where m = mass of gas aspirated.

Also,

$$\frac{P_1 V_1}{T_1} = \frac{P_2 V_2}{T_2}$$

and

$$P_1 V_1{}^n = P_2 V_2{}^n$$

from which

$$\frac{T_2}{T_1} = \frac{P_2}{P_1} \frac{V_2}{V_1}$$

$$= \frac{P_2}{P_1} \left(\frac{P_1}{P_2}\right)^{1/n}$$

$$= \left(\frac{P_2}{P_1}\right)^{(n-1)/n}$$

Substituting in equation (10.15) gives

$$\text{work energy required per cycle} = \frac{n}{n-1}\, mRT_1\left(1 - \frac{T_2}{T_1}\right)$$

$$= \frac{n}{n-1}\, mR(T_1 - T_2) \qquad (10.16)$$

Comparing this with equation (10.3) it can be seen that under similar conditions, the work energy required per pound of gas is not affected by compressor clearance.

10.9 True Volumetric Efficiency

This is defined as the ratio

$$\frac{\text{volume of free air delivered per stroke}}{\text{swept volume}} \qquad (10.17)$$

and will be less than unity owing to

    (a) The pressure drop caused by the resistance of the valves and piping
    (b) The air expanding on entering the hot cylinder

(c) The high pressure air trapped in the clearance volume at the end of the delivery stroke expanding to a pressure just below the suction pressure before the suction valve can open

(d) Leakage of air past the piston

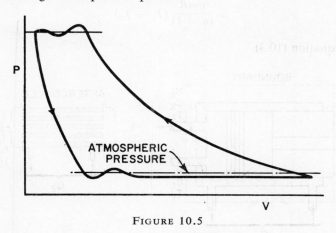

FIGURE 10.5

The true volumetric efficiency will also differ from the diagram volumetric efficiency since the latter is an ideal. The actual indicator diagram will be as in Fig. 10.5, the waves being due to oscillation in the valve opening.

10.10 Rejection of Heat Energy

Compressors are often supplied with jacket cooling water and may also sometimes be fitted with an aftercooler. The purpose of the aftercooler is to cool the compressed air, thus bringing about a decrease in the specific volume of the compressed air, and also causing some of the water vapour held in the air to be condensed and deposited in the aftercooler. This moisture would have a high nuisance value during the subsequent use of the air, as it may well freeze at the temperatures encountered after re-expansion in air motors, etc.

The heat energy rejected to the water jacket may be found by applying the steady flow energy equation between the inlet and outlet of the compressor. The steady flow energy equation gives

$$Q - \frac{W}{J} = m\left[\frac{c_2^2 - c_1^2}{2gJ} + h_2 - h_1\right]$$

The kinetic energy terms may be neglected since their difference will be small compared with the other terms.

For a perfect gas $H_2 - H_1 = m(h_2 - h_1) = mc_p(T_2 - T_1)$ (see chapter 5, §5.6).

Work energy required for m lb of gas

$$= \frac{mnR}{(n-1)}(T_1 - T_2)$$

from equation (10.3).

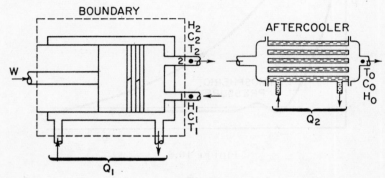

FIGURE 10.6

Substituting in the steady flow equation gives

$$Q_1 - \frac{mnR}{J(n-1)}(T_1 - T_2) = mc_p(T_2 - T_1)$$

$$\therefore \quad Q_1 = m\frac{n}{n-1}\frac{R}{J}(T_1 - T_2) + mc_p(T_2 - T_1)$$

$$= m(T_1 - T_2)\left(\frac{n}{n-1}\frac{R}{J} - c_p\right)$$

$$= \frac{m(T_1 - T_2)}{n-1}\{n(c_p - c_v) - c_p(n-1)\} \text{ since } \frac{R}{J} = c_p - c_v$$

$$= m\frac{(T_1 - T_2)}{n-1}(c_p - nc_v)$$

$$= mC_v(T_1 - T_2)\left\{\frac{\gamma - n}{n-1}\right\} \tag{10.18}$$

This will have a negative value, showing that heat energy is rejected. Similarly, the steady flow energy equation may be applied across the aftercooler. Assuming the air is cooled to its original temperature T_1 in the aftercooler, then referring to Fig. 10.7, $T_0 = T_1$.

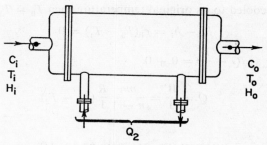

FIGURE 10.7

Applying the steady flow energy equation gives

$$Q - \frac{W}{J} = m\left[\frac{c_0{}^2 - c_i{}^2}{2gJ} + h_0 - h_i\right]$$

Now $c_i \simeq c_0$; $T_i = T_2$ and $T_0 = T_1$; also $W/J = 0$ i.e.,

$$Q_2 - 0 = 0 + mc_p(T_1 - T_2)$$

for a mass flow of m lb

$$\therefore \quad Q_2 = mc_p(T_1 - T_2) \qquad (10.19)$$

$\therefore$ total heat energy interchange $= Q_1 + Q_2$

$$= mc_v(T_1 - T_2)\frac{(\gamma - n)}{n - 1} + mc_p(T_1 - T_2)$$

$$= m(T_1 - T_2)\frac{\gamma c_v - nc_v}{n - 1} + mc_p(T_1 - T_2)$$

$$= m(T_1 - T_2)\left(\frac{c_p - nc_v}{n - 1} + c_p\right)$$

$$= m(T_1 - T_2)\frac{c_p - cn_v + nc_p - c_p}{n - 1}$$

$$= m(T_1 - T_2)\frac{n}{n - 1}(c_p - c_v)$$

$$= \frac{mn}{n - 1}\frac{R}{J}(T_1 - T_2) \qquad (10.20)$$

$= $ work energy transfer in the cylinder

A similar result could be obtained direct by applying the steady flow energy equation across the combined compressor and aftercooler.

The steady flow energy equation gives

$$Q - \frac{W}{J} = m\left[\frac{c_0{}^2 - c_1{}^2}{2gJ} + h_0 - h_1\right]$$

If the air is cooled to its original temperature then $T_0 = T_1$ and

$$h_0 - h_1 = c_p(T_0 - T_1) = 0$$

$$\therefore \qquad Q - \frac{W}{J} = 0 + 0$$

$$\therefore \qquad Q = \frac{W}{J} = \frac{mn}{n-1}\frac{R}{J}(T_1 - T_2) \qquad (10.21)$$

EXERCISES ON CHAPTER 10

All pressures are absolute. For air, take $R = 53\cdot3$ ft lbf/lb degR, $c_p = 0\cdot238$ Btu/lb degR.

1. A single cylinder air compressor takes in air at $14\cdot8$ lbf/in^2 and 60°F, compresses it according to the law $PV^{1\cdot25} =$ constant, and delivers it at 95 lbf/in^2. If clearance may be neglected, calculate the delivery temperature of the air and the isothermal efficiency of the compressor.

(755°R; $82\cdot5\%$)

2. A simple air compressor takes in 75 ft^3/min of air at $14\cdot5$ lbf/in^2 and 60°F, compresses it according to the law $PV^{1\cdot3} =$ constant, and delivers it at 100 lbf/in^2. Neglecting clearance, determine the work done per minute, the mean effective pressure, and the heat transfer to or from the air during compression only, stating the direction of flow.

(382,000 ft lbf; $35\cdot7$ lbf/in^2; -98 Btu/min)

3. An air compressor takes in 30 ft^3/min of air at $14\cdot6$ lbf/in^2 and 70°F and delivers it at 60 lbf/in^2. The equation of the compression curve is $PV^{1\cdot3} =$ constant, and the effect of clearance may be neglected. Determine the air horsepower, the isothermal horsepower, the isothermal efficiency, and the shaft horsepower, assuming a mechanical efficiency of 85%.

($3\cdot22$ hp; $2\cdot71$ hp; $84\cdot2\%$; $3\cdot79$ hp)

4. A single-cylinder single-acting air compressor with negligible clearance delivers air at the rate of 20 lb/min. Suction conditions are $14\cdot5$ lbf/in^2 and 65°F. It may be assumed that during the compression process, 12% of the energy supplied to the air is rejected as heat energy through the cylinder walls. Determine the index of compression and the delivery temperature of the air. If the air is to be cooled at constant pressure to the original temperature in an aftercooler, determine the temperature rise of the aftercooler circulating water if the flow rate of this water is 40 lb/min.

($1\cdot355$; 870°R; 41°degF)

5. Carbon dioxide gas is compressed at the rate of 20 lb/min from 50 lbf/in^2 and 100°F to a pressure of 1,000 lbf/in^2 according to the law $PV^{1\cdot25} =$ constant. Neglecting clearance, determine the required cylinder volume if the compressor is single acting and runs at 300 rev/min, the temperature of the gas after compression, and the work done per minute. Take $R_0 = 1,545$ ft lbf/mol degR.

(315 in^3; 1,020°R; $1\cdot614 \times 10^6$ ft lbf)

6. The clearance volume of an air compressor is 6% of the swept volume. The pressure and temperature of the air during the suction stroke are 14 lbf/in^2

and 90°F respectively, and the delivery pressure is 90 lbf/in². The compression and re-expansion curves follow the law $PV^{1.25}$ = constant, and atmospheric air conditions are 14·7 lbf/in² and 60°F. Determine

(a) The diagram volumetric efficiency
(b) The volumetric efficiency referred to atmospheric conditions
(c) The work done per lb of air

(79·36%; 71·4%; 64,900 ft lbf)

7. A single-cylinder air compressor is to deliver 60 ft³ of free air per minute. The suction conditions are 14·3 lbf/in² and 80°F. The delivery pressure is 120 lbf/in² and compression and re-expansion curves may be assumed to follow the law $PV^{1.25}$ = constant. If the volumetric efficiency of the compressor is 75%, and the mechanical efficiency is 80%, calculate the swept volume in ft³/min and the power required to drive the compressor. Free air conditions are measured at 14·7 lbf/in² and 60°F.

(85·3 ft³/min; 13·2 hp)

8. The clearance volume of a single-cylinder single-acting air compressor is 10% of the swept volume. The compressor has a bore of 4·71 in, a stroke of 6 in, and runs at 400 rev/min. Suction conditions are 14·5 lbf/in² and 75°F, and the delivery pressure is 145 lbf/in². Compression and re-expansion curves follow the law $PV^{1.3}$ = constant. Calculate the diagram volumetric efficiency, the rate of air delivery, and the power input required to drive the compressor, assuming a mechanical efficiency of 74%.

(51·1%; 0·9 lb/min; 3·22 hp)

9. A single-stage, double-acting air compressor takes in air at 14 lbf/in² and 80°F and delivers it at 95 lbf/in². The clearance volume is 5% of the swept volume, and compression and re-expansion curves follow the law $PV^{1.31}$ = constant. The effect of the piston rod area can be neglected and the compressor runs at 450 rev/min. If the compressor delivers 600 ft³/min of free air (measured at 14·7 lbf/in² and 60°F) calculate the bore and stroke of the compressor, assuming a bore to stroke ratio of 1:1·4.

(bore 11·1 in; stroke 15·55 in)

11: Engine Trials

Trials on engines are made in order to determine the actual performance of an engine. Over the years such trials have resulted in improvements in efficiencies with the subsequent lowering of the cost of power, and there is no doubt that the interest taken in the economics of power production has been a considerable incentive in the development of engines. A notable example in this respect is the continuing development of the steam turbine.

In a trial on a heat engine the following are the main results determined:

  (i) The power developed
 (ii) The fuel supplied per unit time or the energy supplied per unit time
(iii) The distribution of the energy supplied
 (iv) The various efficiencies of the engine

11.1 The Power Developed

The power developed at the output shaft of an engine is measured by means of a brake or dynamometer and is expressed as brake horsepower or bhp. In the case of reciprocating engines the power developed in the cylinder is measured by means of an indicator, and expressed in the form of indicated horsepower or ihp. The difference between the ihp and bhp gives a measure of the power dissipated in friction and is termed the friction horsepower or fhp. In steam turbines, the power developed at the output shaft only is measured; this, of course, is the measure of the useful power.

11.2 Measurement of bhp

The rope brake
With reference to Fig. 11.1,

$$\text{effective torque on flywheel } T = W\frac{(D+d)}{2} - S\frac{(D+d)}{2}$$

$$= (W - S)\frac{(D+d)}{2}$$

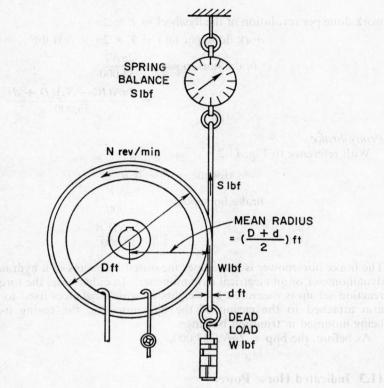

SPRING
BALANCE
S lbf

N rev/min

S lbf

MEAN RADIUS
$= (\dfrac{D + d}{2})$ ft

D ft

d ft

W lbf

DEAD
LOAD
W lbf

FIGURE 11.1

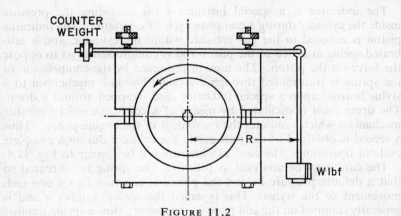

COUNTER
WEIGHT

R

W lbf

FIGURE 11.2

work done per revolution of the flywheel $= T \times 2\pi$

$$\text{work done per min} = T \times 2\pi \times N \text{ ft lbf}$$

$$\text{brake horsepower} = \frac{2\pi NT}{33,000}$$

$$= \frac{2\pi N(W - S)\frac{1}{2}(D + d)}{33,000}$$

Prony brake
With reference to Fig. 11.2

$$\text{torque} = T = WR$$

$$\text{brake horsepower} = \frac{2\pi NT}{33,000}$$

$$= \frac{2\pi NWR}{33,000}$$

The brake horsepower is sometimes measured by means of a hydraulic dynamometer or an electrical dynamometer. In either case, the torque reaction set up is measured by weights or spring balances fixed to an arm attached to the casing of the dynamometer, the casing itself being mounted in trunnion bearings.

As before, the bhp $= 2\pi NT/33,000$.

11.3 Indicated Horse Power

Indicator
The indicator is a special instrument for recording the pressure inside the cylinder during a complete cycle of operation. The indicator piston is exposed to the gas pressure within the cylinder, and a calibrated spring attached to the piston rod is compressed so as to oppose the force on the piston. The motion produced by the compression of the spring is transmitted through a straight-line link mechanism to a stylus bearing upon a special indicator card wrapped around a drum. The drum itself is oscillated by means of a cord attached to another mechanism which simulates the movement of the engine piston. Thus a record is obtained of the pressure in the cylinder during a complete cycle of operation. The card so obtained will be similar to Fig. 11.4.

The card may be analysed as follows. The spring is calibrated so that a definite pressure behind the piston is equivalent to a one inch movement of the stylus. This is called the spring number s and is generally stamped on the end plate of the spring; thus a spring number might be, for example, 90 lbf/in²/in.

Now if the nett area of the diagram, i.e. (+ve area) − (−ve area) is

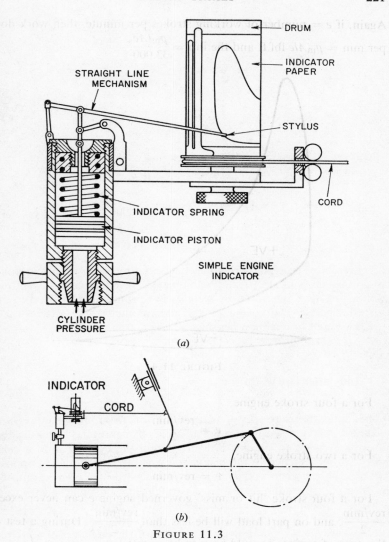

FIGURE 11.3

determined and divided by the length, the mean height h_m is obtained. Then, the indicated mean effective pressure $p_m = h_m \times s$ lbf/in². The mean force behind the engine piston $= p_m A (\text{lbf/in}^2 \times \text{in}^2) = p_m A$ lbf, where A = cross sectional area of piston.

If l = length of the stroke in feet, then

$$\text{work done per cycle} = p_m A l \text{ lbf ft}$$

Again, if e = number of working strokes per minute, then work done per min = $p_m Ale$ lbf ft and the ihp = $\dfrac{p_m lAe}{33,000}$.

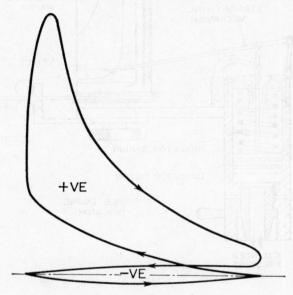

For a four stroke engine,

$$e = \frac{\text{rev/min}}{2}$$

For a two stroke engine,

$$e = \text{rev/min}$$

For a four stroke 'hit or miss' governed engine e can never exceed $\dfrac{\text{rev/min}}{2}$, and on part load will be less than $\dfrac{\text{rev/min}}{2}$. During a test on this type of engine, e would have to be counted.

11.4 Brake mean effective Pressure P_{mb}

This is a hypothetical quantity obtained by using the equation

$$\text{bhp} = \frac{P_{mb} lAe}{33,000}$$

11.5 Mechanical Efficiency

Mechanical efficiency is defined as

$$\frac{bhp}{ihp}$$

Since ihp − bhp is a measure of power dissipated in friction, it is obvious that the higher the value of the mechanical efficiency then the lower is the loss in friction. Mechanical efficiency is given the symbol η_m.

11.6 Fuel Consumption

The amount of fuel consumed by an engine in a given time is an important factor and is always measured during a trial. The consumption depends upon the power developed and the engine efficiency. In the case of reciprocating engines the fuel is, of course, gas, petrol or oil; for gas turbines, various types of oils and sometimes pulverised coal; for steam engines or turbines the amount of steam is measured.

11.7 Consumption Rate or Specific Fuel Consumption

The specific fuel consumption is defined as the consumption of fuel per unit time per unit of power developed. This quantity is much more definite and enables comparisons of similar engines to be made. An engine with a lower specific fuel consumption is the more efficient engine. There are two values of the specific fuel consumption, one based on the ihp (C_i) and the other on the bhp (C_b).

Ex. 11.1. An oil engine consumes 30 lb oil per hour when developing 40 bhp. Find the specific fuel consumption on (i) bhp basis, (ii) ihp basis. Mechanical efficiency = 80%.

$$\text{Specific fuel consumption on b.h.p. basis} = \frac{30}{40}$$

$$= 0.75 \text{ lb/bhp h}$$

$$ihp = \frac{40}{0.8}$$

$$= 50$$

$$\text{Specific fuel consumption on ihp basis} = \frac{30}{50}$$

$$= 0.6 \text{ lb/ihp h}$$

9

Ex. 11.2. A steam turbine has a steam consumption rate of 8·4 lb/kWh. Determine the steam consumption when the turbine generates the full load of 60,000 kW.

$$\text{Steam consumption} = 8\cdot4 \times 60,000$$
$$= 504,000 \text{ lb/h}$$

11.8 Thermal Efficiency

The thermal efficiency of an engine gives a measure of how effectively the heat energy supplied is utilised.

$$\text{The indicated thermal efficiency} = \frac{\text{heat equivalent of the ihp per hour}}{\text{heat supplied per hour}}$$

$$\text{Now heat equivalent of the ihp} = \frac{60 \times 33,000 \times \text{ihp}}{778} \text{ Btu/h}$$

$$= 2,545 \times \text{i.h.p. Btu/h}$$

Heat energy supplied per hour

$$E = (\text{weight of fuel/h}) \times \text{calorific value (C.V.)}$$

or in the case of a gas engine

$$E = (\text{volume of gas/h}) \times \text{calorific value/unit volume}$$

∴ indicated thermal efficiency,

$$\eta_i = \frac{2,545 \text{ ihp}}{(\text{weight of fuel/h}) \times (\text{C.V.})}$$

$$= \frac{2,545}{\{(\text{weight of fuel/h})/\text{ihp}\}(\text{C.V.})}$$

$$= \frac{2,545}{(\text{specific fuel consumption}) \times (\text{C.V.})}$$
$$\text{on ihp basis}$$

$$= \frac{2,545}{C_i \times (\text{C.V.})}$$

Similarly, the brake thermal efficiency,

$$\eta_b = \frac{\text{heat equivalent of the bhp/h}}{\text{heat supplied/h}}$$

$$= \frac{2,545 \times \text{bhp}}{(\text{weight of fuel/h}) \times (\text{C.V.})}$$

$$= \frac{2,545}{\{(\text{weight of fuel/h})/\text{bhp}\} \times (\text{C.V.})}$$

$$= \frac{2,545}{C_b \times (\text{C.V.})}$$

Obviously,

$$\frac{\eta_b}{\eta_i} = \eta_m \text{ (mechanical efficiency)}$$

In the case of a steam engine or steam turbine the calorific value will be replaced by the heat energy available per lb of steam. If h_1 is the enthalpy of the steam at the stop valve of the engine and h_f is the enthalpy of the feed water supplied to the boiler then the heat energy supplied per lb of steam is $(h_1 - h_f)$ and the brake thermal efficiency of the steam engine is equal to $2,545/C_b(h_1 - h_f)$. This efficiency covers all losses including those in the condenser, pipes, bearings and radiation to surroundings.

11.9 Standard Efficiency, η_s

For gas, oil or petrol engines the standard thermal efficiency is taken as the air standard.

$$\eta_s = 1 - \frac{1}{r^{\gamma-1}}$$

(see §8.1).

For steam engines or turbines the standard efficiency is the Rankine Cycle efficiency (see §7.4).

11.10 Efficiency Ratio or Relative Efficiency, η_e

The ratio of the brake thermal efficiency to the standard efficiency is termed efficiency ratio

$$\eta_e = \frac{\text{brake thermal efficiency}}{\text{standard efficiency}} = \frac{\eta_b}{\eta_s}$$

11.11 Energy Balance

The purpose of the energy balance of an engine or plant is to determine the distribution of the energy through the engine or plant. It is usual to express the various items in a balance in terms of a unit time of 1 minute or 1 hour or per unit mass of 1 lb of the fuel or steam supplied. The datum temperature of 32°F or of engine room temperature is often used.

For a reciprocating internal combustion engine Fig. 11.5 shows the apportionment of the energy supplied.

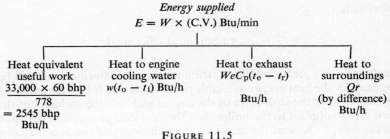

FIGURE 11.5

This distribution is illustrated in Fig. 11.6.

A typical balance sheet for such an engine would be:

	Btu/min	%
Heat energy supplied	2,500	100
Heat equivalent of bhp*	600	24
Heat to engine cooling water	820	32·8
Heat loss to engine exhaust	710	28·4
Heat loss to surroundings (by difference)	370	14·8
	2,500	100

* Some text books include ihp here; this is not to be recommended as part of the ihp will be dissipated in friction which will be transmitted to the cooling water as heat. No balance should include the same items twice.

The energy balance for a steam engine or turbine is different from the above owing to the absence of jacket cooling water and to the return of the working fluid as condensate. A diagrammatic view of a steam plant is given in Fig. 11.7. A datum of 32°F is taken to coincide with the zero of enthalpy. A typical energy balance sheet will be as follows:

	Btu/min	%
Heat energy supplied $= Wh_1$	6,200	100
Heat equivalent of bhp $= \dfrac{33,000}{778}$ bhp	496	8
Heat to condenser cooling water $= W_c(h_0 - h_1)$	4,836	78
Heat in feed water Wh_f	372	6
Heat loss to surroundings (by difference) Q_s	496	8
	6,200	100

Strictly speaking a certain amount of energy is supplied to the extraction pump but normally this is a very small fraction of the useful output and may be neglected at this stage.

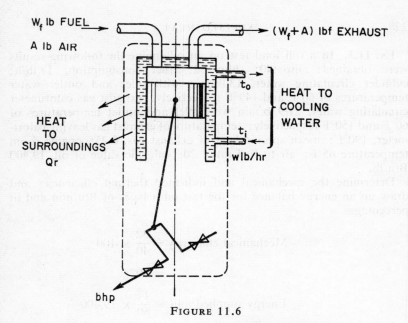

FIGURE 11.6

FIGURE 11.7

Ex. 11.3. In a full load test on an oil engine the following results were obtained: ihp 40; bhp 32; fuel consumption, 17 lb/h; cylinder circulating water, 13 lb/min with inlet and outlet water temperatures of 60°F and 149°F respectively; exhaust gas calorimeter circulating water, 21·9 lb/min with inlet and outlet temperatures of 60°F and 150°F respectively; temperature of exhaust gas leaving calorimeter, 190°F; mean specific heat of exhaust gas, 0·25; engine room temperature 65°F; air to fuel ratio, 20; calorific value of oil, 19,400 Btu/lb.

Determine the mechanical and indicated thermal efficiencies and draw up an energy balance for the test on a basis of Btu/min and in percentages.

$$\text{Mechanical efficiency} = \frac{32}{40} \times 100$$

$$= 80\%$$

$$\text{Energy supplied/min} = \frac{17}{60} \times 19,400$$

$$= 5,500 \text{ Btu}$$

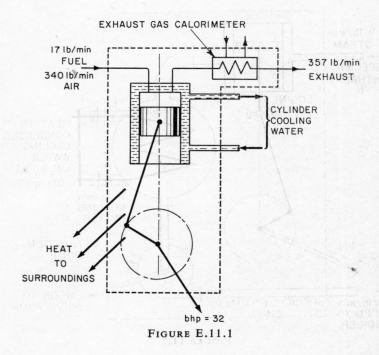

FIGURE E.11.1

$$\text{Heat equivalent of ihp} = \frac{40 \times 33,000}{778}$$

$$= 1,698 \text{ Btu/min}$$

$$\therefore \quad \text{Indicated thermal efficiency} = \frac{1,698}{5,500} \times 100$$

$$= 30 \cdot 9\%$$

$$\text{Heat to engine cooling water} = 13(149 - 60)$$

$$= 1,157 \text{ Btu/min}$$

$$\text{Heat to exhaust gas calorimeter} = 21 \cdot 9(150 - 60)$$

$$= 1,971 \text{ Btu/min}$$

$$\text{Mass of exhaust gas per min} = \frac{(20 + 1) \times 17}{60} = 5 \cdot 95 \text{ lb}$$

Heat in gas leaving calorimeter (above) $= 5 \cdot 95 \times 0 \cdot 25(190 - 65)$

engine room temperature $\qquad = 186 \text{ Btu/min}$

$$\text{Total heat energy in exhaust gas} = 1,971 + 186$$

$$= 2,157 \text{ Btu/min}$$

$$\text{Heat equivalent of bhp} = \frac{32 \times 33,000}{778}$$

$$= 1,358 \text{ Btu/min}$$

Energy Balance

	Btu/min	%
Energy supplied	5,500	100
Heat equivalent of bhp	1,358	24·7
Heat to engine cooling water	1,157	21·0
Heat to exhaust gas	2,157	39·2
Heat to surroundings (by difference)	828	15·1
	5,500	100·0

Ex. 11.4. The following data were obtained in a trial on a single-cylinder double-acting steam engine:

Bore, 6·5 in; stroke, 6 in; piston rod diameter, $1\frac{13}{16}$ in; rev/min, 167; nett brake load, 58 lbf; mean brake diameter, 5·04 ft; mean indicator card area, 1·33 in²; card length, 3 in; spring number, 100 lbf/in²/in; chest pressure and temperature, 71 lbf/in² gauge and 410°F respectively; weight of condensate, 4·2 lbf/min; temperature of condensate leaving condenser, 105°F; weight of cooling water to condenser, 146 lbf/min; cooling water inlet and outlet temperatures, 50°F and 80°F respectively. Atmospheric pressure, 29·6 inHg.

Determine: (a) the bhp, (b) ihp, (c) mechanical efficiency and (d) the thermal efficiency.

Also, draw up an energy balance sheet in Btu/min and in percentages.

$$\text{Mean effective pressure} = \frac{1·33}{3} \times 100$$

$$= 44·4 \text{ lbf/in}^2$$

$$\text{ihp} = \frac{44·4\{\frac{1}{4}\pi 6^2 + \frac{1}{4}\pi\,[6^2 - (1\frac{13}{16})^2]\} \times \frac{6}{12} \times 167}{33,000}$$

$$= \frac{44·4[28·28 + (28·28 - 2·58)] \times 0·5 \times 167}{33,000}$$

$$= 6·0$$

$$\text{bhp} = \frac{2\pi NT}{33,000}$$

$$= \frac{6·28 \times 167 \times 58 \times \frac{1}{2} \times 5·04}{33,000}$$

$$= 4·64$$

$$\text{Mechanical efficiency} = \frac{4·64}{6·00} \times 100 = 77·3\%$$

Energy supply

$$\text{atmospheric pressure} = \frac{29·6}{30} \times 14·7$$

$$= 14·5 \text{ lbf/in}^2$$

$$\text{absolute steam pressure} = 71 + 14·5$$

$$= 85·5 \text{ lbf/in}^2$$

Interpolation from steam tables. Values of enthalpy

Pressure (lbf/in²)	400°F	450°F		
80	1,230·7	1,256·1		
100	1,227·6	1,253·7		
Difference of 20	3·1	2·4		
Difference of 5·5	0·85	0·66	Difference of 50 degF	Difference of 10 degF
∴ at 85·5	1,229·85	1,255·44	25·59	5·12

∴ at 85·5 lbf/in² and 410°F, enthalpy = 1,229·85 + 5·12
$$= 1,235 \text{ Btu/lb}$$

Energy balance above 32°F

	Btu/min	%
Energy supplied = 4·2 × 1,235	5,187	100
Heat equivalent of bhp $= \dfrac{4·64 \times 33,000}{778}$	197	3·8
Heat to cooling water = 146 (80 − 50)	4,380	84·5
Heat in condensate = 4·2 × 73	306	5·9
Heat to surroundings (by difference)	304	5·8
	5,187	100·0

$$\text{Brake thermal efficiency} = \frac{197}{5,187 - 306} \times 100$$

$$= 4·04\%$$

11.12 Typical Graphical Representation of Test Results

In general there are two distinct types of tests: (*a*) on a constant speed engine; (*b*) on a variable speed engine.

The constant speed engine

The test results are represented on a graph having a base of power developed in bhp and are shown in Figs. 11.8 and 11.9.

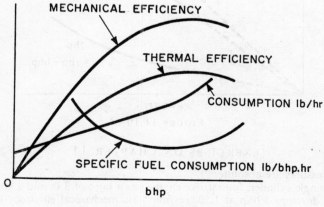

FIGURE 11.8

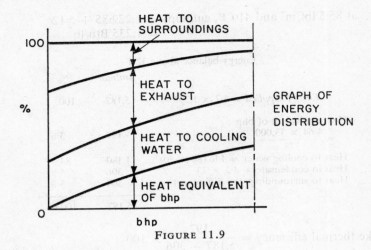

FIGURE 11.9

The variable speed engine

The results are plotted against speed (Figs. 11.10 and 11.11). The test results shown are those appertaining to a certain throttle opening. Tests would normally be carried out for a range of throttle openings.

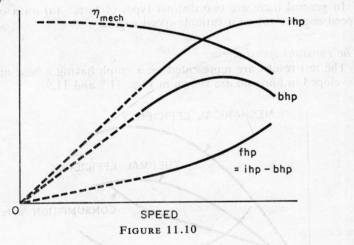

FIGURE 11.10

EXERCISES ON CHAPTER 11

All pressures are absolute.

1. A single-cylinder, four stroke engine has a bore of 3 in and a stroke of 5 in. It develops 5 bhp at 1,500 rev/min. Its mechanical efficiency is 85% and it consumes 2·4 lb of oil per hour of calorific value 18,000 Btu/lb.

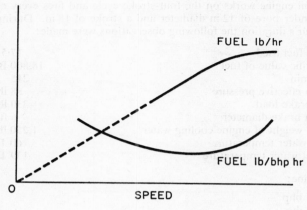

FIGURE 11.11

Determine:

(a) The indicated mean effective pressure
(b) The brake thermal efficiency

and

(c) The fuel consumption in lb/bph h

(87·6 lbf/in², 29·4%, 0·48 lb/bhp h)

2. A four-cylinder, four stroke petrol engine develops 10·6 b.h.p. at 2,000 rev/min. The bore is $2\frac{1}{2}$ in, the stroke 3 in, and the mean effective pressure is 87 lbf/in². Eight pints of petrol are consumed per hour. The specific gravity of the petrol is 0·75 and the calorific value is 18,000 Btu/lb.

Determine:

(a) The mechanical efficiency
(b) The brake thermal efficiency

and

(c) The indicated thermal efficiency

(82%, 25%, 30·5%)

3. A steam turbine develops 10,000 kW when on full load. The steam pressure and temperature at the turbine stop valve are 600 lbf/in² and 800°F, and the temperature of the feed to the boiler is 325°F. The total steam consumption is measured as 86,000 lb/h.

Calculate:

(a) The heat consumption in Btu/kW h

and

(b) The thermal efficiency of the plant

(9,560 Btu/kW h, 27·45%)

4. An oil engine works on the four-stroke cycle and fires every cycle. It has a cylinder bore of 12 in diameter and a stroke of 18 in. During a trial of one hour's duration the following observations were made:

Total fuel used	17·5 lb
Calorific value of fuel	18,800 Btu/lb
Rev/min	215
Mean effective pressure	88 lbf/in²
Net brake load	330 lbf
Mean brake diameter	6 ft 1 in
Total weight of engine cooling water	1,250 lbf
Inlet water temperature	60°F
Outlet water temperature	139°F

Determine:

(a) The bhp
(b) The ihp

and

(c) The brake thermal efficiency
Also, draw up an energy balance for the trial in Btu/min and in percentages.
(41·1, 48·7, 31·8%. Energy balance: heat equivalent of bhp 1,742 Btu/min, 31·8%; heat to cooling water, 1,645 Btu/min, 30%; heat to exhaust and to surroundings, 2,097 Btu/min, 38·2%)

5. During a test on a single cylinder double-acting steam engine the following results were obtained:

Mean effective pressure	50 lbf/in²
Engine speed	152 rev/min
Net brake load	225 lbf
Effective brake diameter	5 ft
Steam consumption rate	500 lb/h
Steam supply pressure	200 lbf/in²
Dryness fraction of steam supply	0·95
Condenser circulating water rate	65 lb/min
Temperature rise of circulating water	75 degF
Condensate temperature	120°F

If the engine has a bore of 8 in and a stroke of 12 in, calculate the ihp, bhp, mechanical efficiency, and specific steam consumption (in lb/bhp h). Draw up a percentage energy distribution from the above data.
(23·15; 16·28; 70·3%; 30·75 lb/bhp h)
(as percentage of energy input; bhp = 7·18%, condenser circulating water 50·7%, condensate 7·62%, radiation etc., 34·5%)

6. The following data were obtained during a trial on a single-cylinder four stroke oil engine of 8 in bore and 13 in stroke.

Engine speed	400 rev/min
Torque reading on dynamometer	230 lbf ft
Fuel consumption rate	9·1 lb/h

Air consumption rate	250 lb/h
Calorific value of fuel	18,200 Btu/lb
Engine exhaust temperature	830°F
Mean specific heat of exhaust gases	0·26 Btu/lb degR
Engine cooling water flow rate	10 lb/min
Engine cooling water temperature rise	66 degF
Mean effective pressure	87 lbf/in²
Laboratory temperature	70°F

Calculate the mechanical efficiency, brake thermal efficiency, and specific fuel consumption (in lb/bhp h). Draw up an energy balance using laboratory temperature as datum.

(61%; 26·9%; 0·52 lb/bhp h)
(as percentage of energy input; bhp = 26·9%; cooling water = 23·9%; exhaust gas = 30·85%; radiation, etc. = 18·35%.)

7. A test on a single-cylinder four stroke gas engine governed by a 'hit and miss' governor provided the following results.

Net brake load	320 lbf
Brake diameter	4 ft
Average engine speed	215 rev/min
Average explosions per minute	100
Indicator card area	0·645 in²
Indicator card length	2·2 in.
Indicator spring rating	300 lbf/in² per in
Gas consumption rate	670 ft³/h
Calorific value of gas used	450 Btu/ft³
Cooling water flow rate	1,200 lb/h
Cooling water temperature rise	86 degF
Cylinder bore	12 in
Stroke	15·25 in

Calculate the brake thermal and mechanical efficiencies, and draw up a percentage energy balance as far as is possible.
(22·1%; 68·4%; as percentage of heat energy input; bhp = 22·1%; energy to cooling water = 34·3%; energy to exhaust, radiation, etc. = 43·6%)

8. A test conducted on a four-stroke, four cylinder petrol engine having a bore of 3 in and a stroke of 4 in produced the following results:

Engine speed	3,000 rev/min
Dynamometer torque	52 lbf ft
Mean effective pressure per cylinder	106 lbf/in²
Rate of fuel consumption	19·8 lb/h
Calorific value of fuel	18,000 Btu/lb
Cooling water flow rate	23·6 lb/min
Cooling water temperature rise	75 degF

If the clearance volume per cylinder is 3·54 in³, calculate:

(a) The brake horsepower
(b) The indicated horsepower
(c) The mechanical efficiency

(*d*) The brake thermal efficiency
(*e*) The air standard efficiency
(*f*) The relative efficiency
(*g*) The specific fuel consumption (in lb/bhp h)

Draw up a percentage energy balance using the above data.

(29·7; 45·4; 0·655; 0·212; 0·585; 0·358; 0·666.)

As percentage of energy input; bhp = 21·2%, energy to cooling water = 29·8%, energy to exhaust, radiation, etc. = 49%.

Appendix 1

Pressure

Pressure is an important property of a system. It is defined as a force acting upon a unit area of the enclosing boundary. It is caused by the bombardment of the boundary by the molecules of the fluid comprising part of the system. The pressure acting on the walls of a containing vessel is taken as positive and in fact all pressures are positive. If all the fluid is evacuated from a vessel the pressure will be zero. Pressure is measured in units of force/units of area, e.g. lbf/ft^2 or lbf/in^2.

Pressure due to a head of fluid

If the specific weight of a liquid is w lbf/ft^3 then a 1 ft head of liquid will result in a force of w lbf on a 1 ft^2 area.

Hence

$$\text{the pressure for a head of 1 ft} = w \ lbf/ft^2$$

and

$$\text{the pressure for a head of } z \text{ ft} = wz \ lbf/ft^2$$

or $\dfrac{wz}{144}$ lbf/in^2 (1)

For water $w = 62 \cdot 4$ lbf/ft^3

Substituting in (1) gives the pressure for z ft of water

$$= \frac{62 \cdot 4}{144} z \ lbf/in^2$$

$$= 0 \cdot 433 \ z \ lbf/in^2$$

The specific gravity of mercury is $13 \cdot 6$ and hence the pressure due to 1 ft of mercury $= 0 \cdot 433 \times 13 \cdot 6$ lbf/in^2.

The pressure due to 1 in of mercury $= 0 \cdot 433 \times 13 \cdot 6 \times \dfrac{1}{12}$ lbf/in^2

$$= 0 \cdot 491 \ lbf/in^2$$

Barometer

A simple barometer can be made by filling a glass tube having a length longer than 30 in, inverting it, and placing the open end in a mercury bath as shown in the Fig. A1.1. The space at the upper end is almost a perfect vacuum and the column of mercury is supported by

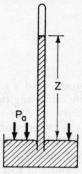

FIGURE A1.1

the air pressure acting on the surface of the mercury in the bath. The head of the mercury, therefore, is a measure of the air pressure.

If the head is z in, then the air pressure

$$P_a = 0.491\ z\ \text{lbf/in}^2$$

If $z = 30$ in, $P_a = 0.491 \times 30 = 14.73\ \text{lbf/in}^2$.

U-tubes

U-tubes containing mercury or any other suitable fluid are used for measuring system pressure.

(i) System pressure above atmostphere—open U-tube containing mercury

$$P_s - P_a = 0.491\ z\ \text{lbf/in}^2$$

$\therefore$

$$P_s = P_a + 0.491\ z\ \text{lbf/in}^2$$

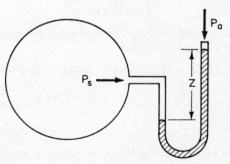

FIGURE A1.2

(ii) System pressure below atmospheric—open U tube containing mercury

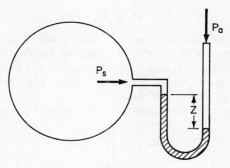

FIGURE A1.3

Here $P_s < P_a$

$\therefore$
$$P_a - P_s = 0.491\,z\ \text{lbf/in}^2$$
$$P_s = P_a - 0.491\,z\ \text{lbf/in}^2$$

Bourdon Gauge

This gauge is used for measuring the difference between the system pressure P_s and the air pressure P_a in cases which are outside the range of U-tubes.

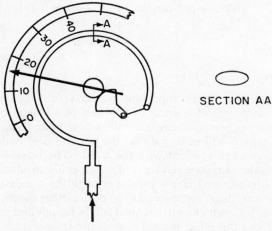

SECTION AA

FIGURE A1.4

The reading is termed the gauge pressure P_G.

$$P_s - P_a = P_G$$

$$\therefore \qquad\qquad P_s = P_G + P_a$$

P_s is termed the absolute pressure. The standard atmospheric pressure is taken as $14 \cdot 73$ lbf/in². A pressure below the atmospheric pressure is termed as 'vacuum', sometimes denoted as so many inches vacuum below atmospheric pressure, e.g. in Fig. A1.3 the system pressure is z in of mercury vacuum.

Fig. A1.5 illustrates the terms used in denoting pressures.

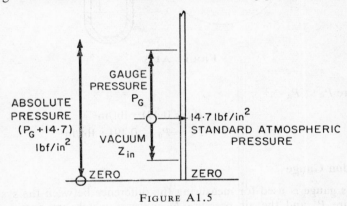

FIGURE A1.5

E.g. if $z = 10$ in mercury column, then

$$\begin{aligned}
\text{absolute pressure} &= 14 \cdot 73 - 10 \times 0 \cdot 491 \\
&= 14 \cdot 73 - 4 \cdot 91 \\
&= 9 \cdot 82 \text{ lbf/in}^2
\end{aligned}$$

Temperature

Temperature is another important property of a system. It is a measure of the degree of hotness or coldness of a system or a body.

If a hot body is placed in contact with a cold body then, after some time has elapsed, it will be found that the hot and cold bodies give the same sensation when touched with the hand. The bodies are said to have attained the same temperature. The sensation produced by a hot body when it is touched is not precise enough for a measure of temperature and it is, therefore, necessary to devise other means for precise measurement. It is well known that most solids, liquids and gases expand when heated and this effect is used to measure temperature. The most common method is by means of the mercury in glass thermometer;

in this case mercury expands or contracts in a fine capillary tube as the temperature rises or falls.

Scales of temperature

There are two scales of temperature in general use: the Fahrenheit and the Centigrade (Celsius) scales. The temperature at which water freezes under a standard atmospheric pressure is termed the ice point; this point is arbitrarily assigned the number 32 on the Fahrenheit scale and 0 on the Centigrade scale. The temperature at which the water boils under a standard atmospheric pressure is given the value 212 on the Fahrenheit scale and 100 on the Centigrade scale. Between these two points the scale is subdivided into 180 equal amounts on the Fahrenheit scale and 100 on the Centigrade scale. It is clear, therefore, that 180 degF ≡ 100 degC or 1 degF ≡ 5/9 degC.

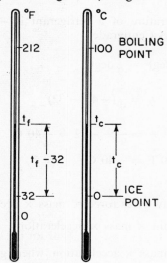

FIGURE A1.6

Now, $$t_c = \frac{5}{9}(t_f - 32) \text{ or } t_f = \frac{9}{5}t_c + 32$$

Absolute Temperature

The absolute temperature is the zero below which the temperature cannot fall. This temperature has never been reached although temperatures of small fractions of a degree above the absolute zero have been reached in some experiments. If 'permanent gases' such as oxygen or nitrogen are used as the substances in a thermometer the absolute zero can be deduced from the rate of expansion or contraction of the gas under constant pressure.

The absolute zero is found to be 273°C or 492°F below the ice point. Since the ice point is at 32°F it must be noted that the absolute zero on the Fahrenheit scale is at −460°F.

Ex. 1. The temperature of steam leaving a boiler is at 490°F. Express this temperature in degrees Centigrade.

$$1 \text{ degF} \equiv \frac{5}{9} \text{ degC}$$

$$t_c = \frac{5}{9}(490 - 32)$$

$$t_c = \frac{5}{9} \times 458$$

$$= 254 \cdot 4°C$$

Ex. 2. The temperature of a refrigerant is −40°F. What is its temperature in degrees Centigrade.

$$1 \text{ degF} \equiv \frac{5}{9} \text{ degC}$$

$$t_c = \frac{5}{9}(-40 - 32)$$

$$= \frac{5}{9} \times -72 = -40°C$$

$$\therefore \quad -40°F = -40°C$$

Unit of Force

Newton's second law gives force $\propto$ mass $\times$ acceleration, i.e.

force = constant $\times$ mass $\times$ acceleration

$$= \frac{1}{g_c} \times \text{mass} \times \text{acceleration, where } g_c \text{ is a constant.}$$

If the unit of mass is taken as 1 lb, then the unit of force (1 lbf) is defined as the gravitational force exerted on a mass of 1 lb at London, where the acceleration due to gravity is approximately 32·2 ft/sec².

$$\therefore \quad 1 \text{ lbf} = \frac{1}{g_c} \times 1 \text{ lb} \times 32 \cdot 2 \text{ ft/sec}^2$$

$$\therefore \quad g_c = 32 \cdot 2 \text{ lb ft/lbf sec}^2$$

Hence the weight (lbf) of a body of mass m (lb) is given by

$$\text{weight (lbf)} = \frac{m \text{ (lb)} \times g \text{ ft/sec}^2}{g_c \text{ lb ft/lbf sec}^2} = \frac{mg}{32 \cdot 2}$$

Appendix 2

Area under various P–V Curves

1. To find the area bounded by

(a) An equation of the form $PV^n = $ constant
(b) The volume axis
(c) The initial and final volume

Consider an element of area dA (Fig. A2.1(b))

$$dA = P \, dV$$

$\therefore$ total area
$$A = \int_{V_1}^{V_2} P \, dV$$

$$= \int_{V_1}^{V_2} \frac{c}{V^n} \, dV \text{ since } PV^n = c$$

$$= c \int_{V_1}^{V_2} V^{-n} \, dV$$

$$= c \left[\frac{V^{1-n}}{1-n} \right]_{V_1}^{V_2}$$

$$= \frac{c}{1-n} [V_2^{1-n} - V_1^{1-n}]$$

$$= \frac{c}{1-n} \left[\frac{V_2}{V_2^{\,n}} - \frac{V_1}{V_1^{\,n}} \right]$$

$$= \frac{1}{1-n} \left[\frac{c}{V_2^{\,n}} V_2 - \frac{c}{V_1^{\,n}} V_1 \right]$$

Since
$$PV^n = c$$

then
$$P_1 V_1^{\,n} = c$$

$\therefore$
$$\frac{c}{V_1^{\,n}} = P_1$$

similarly
$$\frac{c}{V_2^{\,n}} = P_2$$

$\therefore$ total area
$$= \frac{1}{1-n} [P_2 V_2 - P_1 V_1]$$

243

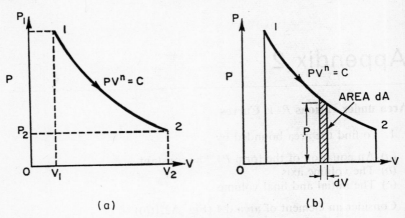

FIGURE A2.1

Since n is usually greater than unity, the expression is normally written:

$$\text{total area} = \frac{P_1 V_1 - P_2 V_2}{n - 1}$$

2. To determine the area bounded by

(a) A curve of the form $PV = \text{constant}$
(b) The volume axis
(c) The initial and final volumes

Consider an element of area dA Fig. A2.2(b)

$$dA = P\,dV$$

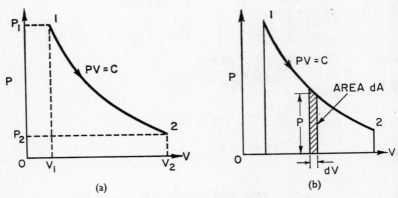

FIGURE A2.2

$$\therefore \quad \text{total area } A = \int_{V_1}^{V_2} P \, dV$$

$$= \int_{V_1}^{V_2} \frac{c}{V} \, dV \text{ since } PV = c$$

$$= c \int_{V_1}^{V_2} \frac{dV}{V}$$

$$= c[\ln V]_{V_1}^{V_2}$$

$$= c \, [\ln V_2 - \ln V_1]$$

$$= c \ln \frac{V_2}{V_1}$$

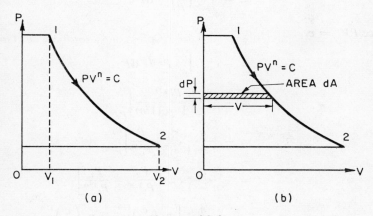

(a)   (b)

FIGURE A2.3

Since $PV = c$ then $P_1 V_1 = c = P_2 V_2$

$$\therefore \quad \text{total area } A = P_1 V_1 \ln \frac{V_2}{V_1}$$

$$= P_1 V_1 \ln \frac{P_1}{P_2}$$

$$= P_2 V_2 \ln \frac{V_2}{V_1}$$

$$= P_2 V_2 \ln \frac{P_1}{P_2}$$

3. To find the area enclosed by

(a) A curve of the form $PV^n = $ constant
(b) The pressure axis
(c) The initial and final pressures

Consider an element of area dA (Fig. A2.3(b))

$$dA = -V\,dP$$

since dP will be negative

$$\therefore \qquad \text{total area } A = -\int_{P_1}^{P_2} V\,dP$$

$$= -\int_{P_1}^{P_2} \left(\frac{c}{P}\right)^{1/n} dP$$

since $PV^n = c$

$$= -c^{1/n}\int_{P_1}^{P_2} P^{-1/n}\,dP$$

$$= -c^{1/n}\left[\frac{P^{1-(1/n)}}{1-(1/n)}\right]_{P_1}^{P_2}$$

$$= -\frac{n}{n-1}\,c^{1/n}\left[\frac{P}{P^{1/n}}\right]_{P_1}^{P_2}$$

$$= -\frac{n}{n-1}\,c^{1/n}\left[\frac{P_2}{P_2^{1/n}} - \frac{P_1}{P_1^{1/n}}\right]$$

$$= -\frac{n}{n-1}\left[P_2\left(\frac{c}{P_2}\right)^{1/n} - P_1\left(\frac{c}{P_1}\right)^{1/n}\right]$$

Since $\qquad PV^n = c$

then $\qquad P_1 V_1{}^n = c$

and $\qquad V_1 = \left(\frac{c}{P_1}\right)^{1/n}$

similarly, $\qquad V_2 = \left(\frac{c}{P_2}\right)^{1/n}$

$$\therefore \qquad \text{total area} = -\frac{n}{n-1}\,[P_2 V_2 - P_1 V_1]$$

$$= \frac{n}{n-1}\,[P_1 V_1 - P_2 V_2]$$

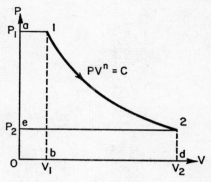

FIGURE A2.4

Alternatively

Area required = area a12e (Fig. A2.4)

$$= \text{area a1b0} + \text{area 12db} - \text{area e2d0}$$

$$= P_1 V_1 + \frac{P_1 V_1 - P_2 V_2}{n-1} - P_2 V_2$$

$$= (P_1 V_1 - P_2 V_2) + \frac{1}{n-1}(P_1 V_1 - P_2 V_2)$$

$$= (P_1 V_1 - P_2 V_2)\left(1 + \frac{1}{n-1}\right)$$

$$= \frac{n}{n-1}(P_1 V_1 - P_2 V_2)$$

If the equation to the curve is $PV = c$, i.e. a hyperbolic curve, then the area a12e = area a1b0 + area 12bd − area e2d0

$$= P_1 V_1 + P_1 V_1 \ln \frac{V_2}{V_1} - P_2 V_2$$

but

$$P_1 V_1 = P_2 V_2$$

$\therefore$

$$\text{area a12e} = P_1 V_1 \ln \frac{V_2}{V_1}$$

or

$$P_1 V_1 \ln \frac{P_1}{P_2}$$

Index